Going Nuclear

FREE PROOF COPY – NOT FOR RESALE

This is an uncorrected book proof made available in confidence to selected persons for specific review purpose and is not for sale or other distribution. Anyone selling or distributing this proof copy will be responsible for any resultant claims relating to any alleged omissions, errors, libel, breach of copyright, privacy rights or otherwise. Any copying, reprinting, sale or other unauthorized distribution or use of this proof copy without the consent of the publisher will be a direct infringement of the publisher's exclusive rights and those involved liable in law accordingly.

Also by Tim Gregory

Meteorite: The Stones From Outer Space That Made Our World

Going Nuclear
How the Atom Will Save the World

TIM GREGORY

THE BODLEY HEAD
LONDON

1 3 5 7 9 10 8 6 4 2

UK | USA | Canada | Ireland | Australia
India | New Zealand | South Africa

The Bodley Head, an imprint of Vintage, is part of the
Penguin Random House group of companies

Vintage, Penguin Random House UK, One Embassy Gardens,
8 Viaduct Gardens, London SW11 7BW

penguin.co.uk/vintage
global.penguinrandomhouse.com

First published by The Bodley Head in 2025

Copyright © Tim Gregory 2025

The moral right of the author has been asserted

The Author is writing the Work in a personal capacity. Any views and opinions expressed in the Work are solely those of the Author and do not represent those of the United Kingdom National Nuclear Laboratory Ltd

Penguin Random House values and supports copyright. Copyright fuels creativity, encourages diverse voices, promotes freedom of expression and supports a vibrant culture. Thank you for purchasing an authorised edition of this book and for respecting intellectual property laws by not reproducing, scanning or distributing any part of it by any means without permission. You are supporting authors and enabling Penguin Random House to continue to publish books for everyone. No part of this book may be used or reproduced in any manner for the purpose of training artificial intelligence technologies or systems. In accordance with Article 4(3) of the DSM Directive 2019/790, Penguin Random House expressly reserves this work from the text and data mining exception.

Set in 12/14.75pt Bembo Book MT Pro
Typeset by Jouve (UK), Milton Keynes
Printed and bound in Great Britain by Clays Ltd, Elcograf S.p.A.

The authorised representative in the EEA is Penguin Random House Ireland, Morrison Chambers, 32 Nassau Street, Dublin D02 YH68

A CIP catalogue record for this book is available from the British Library

HB ISBN 9781847928078
TPB ISBN 9781847928085

Penguin Random House is committed to a sustainable future for our business, our readers and our planet. This book is made from Forest Stewardship Council® certified paper.

For Amy, with whom I walk in Lakeland

I believe all young people think about how they would like their lives to develop; when I did so, I always arrived at the conclusion that life need not be easy provided only it was not empty.

Lise Meitner, *Bulletin of the Atomic Scientists*, 1964

Contents

Acknowledgements	xi
Prologue	1
Going Nuclear: An Introduction	5
CHAPTER 1: WRESTING FIRE FROM THE GODS	7
CHAPTER 2: HOW NUCLEAR POWER WORKS	30
CHAPTER 3: UNRELIABLES	50
CHAPTER 4: NET ZERO IS IMPOSSIBLE WITHOUT NUCLEAR POWER	67
CHAPTER 5: THE GREEN SHEEN	92
CHAPTER 6: NUCLEAR WASTE	105
CHAPTER 7: NUCLEAR FOR THE THIRD MILLENNIUM	124
CHAPTER 8: RADIOPHOBIA	147
CHAPTER 9: WE NEED TO TALK ABOUT CHERNOBYL	164
CHAPTER 10: GOLDEN GEESE	184
CHAPTER 11: RADIOACTIVE REMEDIES	200
CHAPTER 12: STOCKPILES AND SLEUTHS	226
CHAPTER 13: THE FINAL FRONTIER	254
EPILOGUE: APPEALING TO OUR BETTER NATURE	275
Appendix	277
Numerical prefixes	277
List of Illustrations	279
Bibliography	281
Notes	317
Index	347

Acknowledgements

Science doesn't happen in isolation. Nor does writing a book. I have many people to thank for their part in making *Going Nuclear* possible.

First and foremost, I'm deeply thankful to Alice Skinner at The Bodley Head for so carefully editing this book and giving it its shine. Her belief in this project from the first day was a sustaining force as I wrote.

I'm also grateful to my agents at Northbank Talent Management, especially Diane Banks, whose encouragement makes me braver, and Matt Cole – my literary agent – who lifted *Going Nuclear* off the launchpad. Martin Redfern convinced me to put pen to paper for a second time. Sam Wells expertly buffed the manuscript during copyediting.

I'm thankful to my community of colleagues at the United Kingdom National Nuclear Laboratory who supported me as I wrote. Mike Edmondson, Beth Slingsby, Mat Budsworth, Howard Greenwood, and Robin Taylor generously reviewed chapters in their areas of expertise; their prowess and incisive comments sharpened my thinking and strengthened my words. Mike Sloggert, Chris Colton, Steve Shackleford, David McCaw, Laura Leay, and Claire Hindle cleared parts of the manuscript before they were released into the public domain. Keith Franklin expertly reviewed drafts of Chapters 9 and 12, and imparted much wisdom when it comes to politics. Katie Baverstock-Hunt – who coordinated the clearance effort and read an early draft of the entire manuscript – showed me kindness, patience, and encouragement throughout. The Measurement and Analysis team – with whom I rub shoulders in the lab – were, and continue to be, the heart and soul of Central Laboratory. And down the road from the United Kingdom National Nuclear Laboratory, I'm grateful to Matt Legg and Howard Rooms at Sellafield Ltd for supporting me so enthusiastically; they are two of the many unsung heroes of the nuclear industry.

I'm also thankful to the many people who kindly gave me pointers and primers as I meandered my way towards a finished book: Aivija Grundmane helped me with my organic chemistry, Penny Wilson helped me with my hieroglyphs, Meera Alblooshi helped me with my Arabic, and Dr Janey Gregory helped me translate medical jargon into plain English.

I'm grateful for the heartening encouragement of my friends. Lucy Coriander Manifold kept me going when the writing got tough, as she always does. Andy Wrigley made sure I got plenty of fresh air by dragging me up the fells at warp speed. Tim Tinsley, as well as expertly reviewing an early draft of Chapter 13, reminded me of the importance of using my voice to advocate for science's untapped potential. The people who simply asked how the writing was going – especially Sam(antha) Beadsmoore, Richard Booth, Sophie Williams, and Phil Gravett – reminded me I wasn't doing it on my own. And I'm grateful for the purrs from Balthazar, who watched me write much of this book as he lounged on my desk, and for Missy and two torties, who kept me company as I wrote the first part of Chapter 8.

My mother – Janet Gregory – raised me and my sister alone. She taught me many things, not least the importance of hard work and compassion. Thank you, Mum.

And finally, I'm grateful to Amy Gregory – who during my writing of this book became my wife – for seeing it through.

Prologue

The radiation probe starts crackling when I hold it against the small metal disc in my gloved hand. My heart quickens. No need to worry, though. This is exactly what I expected would happen.

Click, click . . . click, click, click . . . click . . .

Most of the samples in the lab are radioactive. Some are less radioactive than a banana. Others are so radioactive that we store them behind walls built from lead bricks. My labmates and I work with these samples every day, but the thrill of detecting their radioactivity never fades.

Click, click, click . . . click . . . click . . . click, click . . .

This morning, I'm repairing the front end of a mass spectrometer, an instrument that splits matter into its constituent atoms like a prism splitting a beam of sunlight into a rainbow. The small metal disc has borne the brunt of the sample I've just measured, and so had become sullied.

Click . . . click . . . click, click . . . click . . .

The radiation makes itself audible by sparking tiny currents in the probe. With each *click*, a radioactive atom has transformed from one element into another. Matter alchemises before me.

Atoms are everywhere, but they're easy to miss. Not in a nuclear lab, though. The crackling tumult of decaying atoms takes you into the nuclear realm at the bottom of material reality. Thinking in terms of atoms turns everyday objects into miniature universes.

Click . . . click . . . click, click, click . . .

I take the radiation probe away from the metal disc. It falls silent. The only noises remaining are the dulcet hums of the fume hood extractor and the whirs of the vacuum pumps. I push the metal disc into the front end of the mass spectrometer and screw it carefully into place. There's no point cleaning it; I've got more samples to measure.

Using the radiation probe, I dutifully frisk every square inch of my gloved hands. No *clicks*. Good, I'm not contaminated. (This isn't surprising; it's never happened to me before.) I take off the gloves by turning them inside-out – careful not to touch the outside with my bare hands, just in case – and drop them into the low-level nuclear waste bin.

For good measure, I frisk my arms and upper body with the radiation probe, too. Another reassuring absence of *clicks*.

I power up the mass spectrometer and start running the digital chart recorder. The trace scrolls as a flat line from left to right. I load a new sample, and as the concoction flows through the various capillaries and conduits, the mass spectrometer springs to life: it splits the sample into atom-sized pieces and beams them into the detector that feeds the chart recorder.

Flat line, flat line, flat line, and then . . . a series of pulses appear, like an ECG tracing out a heartbeat.

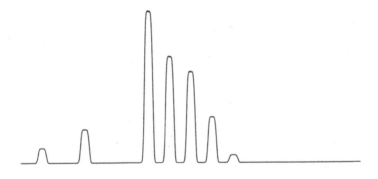

Each pulse marks the presence of a distinct type of atom.

The first beat appears at atomic mass number 233. It's uranium-233, an atom capable of firing the furnace inside a nuclear reactor. Like all the atoms on the chart recorder, uranium-233 is radioactive. Left to its own devices, it spontaneously transforms into new atomic varieties, some of which – when treated with due care and respect – can be turned into an advanced treatment for fatal diseases.

Onwards, the chart recorder scrolls.

The second beat appears at atomic mass number 235. It's uranium-235, the atom that changed the world when humans began orchestrating nuclear chain reactions in the 1940s. In splitting atoms of uranium-235, we acquired the weaponry that catalysed the atomic arms race of the twentieth century. At the same time, we tapped a source of energy so plentiful that it could safely power our energy-hungry civilisation into the far future with little detriment to the natural environment. Uranium-235, more than any other atom, will decide whether or not we successfully renounce fossil fuels and decarbonise our energy systems.

The chart recorder continues to scroll.

The next beat appears at atomic mass number 238. It's uranium-238, the most common form of uranium. It doesn't sustain nuclear chain reactions, but it can be cultivated as a seed from which new types of nuclear fuel grow. Whilst uranium-235 could help us reach net zero by 2050, uranium, uranium-238 could power the world cleanly for a millennium.

Uranium-238 is also the heaviest atom to exist naturally on planet Earth. If this were an everyday material in a normal chemistry lab, the chart recorder would flatline here. But this sample is neither natural nor normal. Some of its atoms saw their origins inside a nuclear reactor.

The next beat on the chart recorder is at atomic mass number 239. It marks the presence of an atom that exists only because of our acquaintance with and profound understanding of the chemical world: plutonium-239.

Plutonium was the first human-made chemical element that we created in quantities large enough to see with the naked eye. Initially, we synthesised plutonium by the millionth of a gram on chemistry benchtops. In time, we forged many thousands of tonnes inside the cores of nuclear reactors. I'm measuring mere billionths of grams in the lab today, but the biggest civil plutonium stockpile in the world – containing hundreds of *tonnes* – lies in Sellafield, where I work. Synthetic elements in such profuse quantities make this site one of the most chemically exotic square miles on the planet.

To make plutonium, we transformed the chemical building blocks of matter through nuclear reactions. We, humans, are not mere features in the chemical world; we are shapers of it. The existence of plutonium is so profound that geologists have proposed using its appearance on Earth in the 1940s to define the beginning of a new epoch: the Anthropocene. Geologically speaking, plutonium may come to define our planetary legacy.

Still onwards, the chart recorder scrolls.

The final traces reveal the rarer and heavier types of plutonium: plutonium-240, plutonium-241, and, finally, plutonium-242.

The penultimate beat, plutonium-241, draws my eye. Some of my colleagues recycle waste plutonium-241 for its radioactive progeny – americium-241 – to make nuclear space batteries. These batteries will someday power spacecraft in the solar system. They might even power human habitats on the Moon, or perhaps Mars. Astronomically speaking, plutonium may come to define our interplanetary legacy.

After plutonium-242, the chart recorder flatlines. This is as far up the periodic table as this sample goes. No heavier atom lies beyond.

I programme the mass spectrometer to measure the next batch of samples. I frisk the rest of my body and the underside of my shoes with the radiation probe, and I leave the lab satisfied. After passing through the security checkpoint guarding the lab entrance, and one last frisk, I head back to my office for a coffee.

Going Nuclear: An Introduction

It feels as though we're in perpetual crisis, especially when it comes to what makes our world go round: energy.

Global warming intensifies as we burn more and more fossil fuels. Our culture is saturated by environmental anxiety. There is an overwhelming mood of foreboding, especially amongst young people. In their bid to renounce fossil fuels, governments are desperately seeking emissions-free sources of energy capable of growing economies and elevating living standards.

Causes for optimism seem scarce. But deep inside the atom, there lies a credible antidote to the despair: *the nucleus*. Brimming with energy like a candle in the dark, the nucleus could power the world securely, reliably, affordably, and – crucially – sustainably. With nuclear power, we can enjoy energy abundance *and* preserve the natural environment.

However, a landmark survey in 2023 quantified what has long been apparent: the most climate-concerned environmentalists are the least supportive of nuclear power. Campaign groups – including Greenpeace, Friends of the Earth, the World Wide Fund for Nature (WWF), 360.org, and the Sierra Club – all oppose nuclear power. And the general public is often suspicious. Opposition to emissions-free energy – at a time when we need it most – is one of the great contradictions of our age.[1]

In the pages that follow, I'll show why this mistrust is misplaced, and how nuclear power is the only way can we decarbonise our world whilst sustaining economic growth, protecting the environment, and continuing the human progress that has characterised the past century. Splitting atoms of uranium inside nuclear reactors is our best shot at reaching net zero by 2050. Beyond the immediate and pressing energy transition, we'll see how we can use nuclear science to recycle radioactive waste and alchemise the chemical elements that could power the world cleanly *for a millennium*.

We'll discover that nuclear science has far more to offer the world than just ('just') clean power, too. We'll meet the mavericks who use radiation to eradicate pests and the atomic gardeners who grow super-crops in gamma gardens. We'll rifle through the radioactive chemistry set that scientists and physicians are turning into novel cures for cancer. We'll encounter the forensic investigators who use nuclear science to solve crimes. And we'll learn how nuclear power enabled us to explore the solar system, and how, soon, it will permit humanity to thrive on the surfaces of other worlds.

At the heart of *Going Nuclear* lies my belief that science and technology – implemented across the globe by our shared institutions – are two of our best tools for promoting human prosperity and protecting the natural world. I wrote these words in the hope they will inspire others to advance those causes, sustainably and responsibly, in a spirit of rational and daring optimism.

Chapter 1. Wresting Fire from the Gods

By convention sweet and by convention bitter;
by convention hot and by convention cold;
by convention colour;
but in reality, atoms and void.

– Democritus, c. 460–370 BC

Imagine we have a grape. Instead of crushing it into wine, we slice it in half with a sharp knife. Now imagine we take one of those halves and slice *that* in half, and again, and so on. After a dozen slices, we'd find ourselves with a piece of grape smaller than a grain of sand.

Now imagine we keep going. After a dozen more slices, our morsel would be the width of a human hair. Another dozen slices and it would be the size of a red blood cell. Seven slices more, and it would be the size of a bacterium.

If we persevered until we'd halved our grape 80 times, we'd encounter the smallest unit of ordinary matter from which our world is made. The ancient Greek philosopher Democritus would have named our impossibly small piece of grape ἄτομος – 'atomos' – meaning, appropriately enough, *uncuttable*. In his spirit, we'd call it by a similar name: an atom.

Atoms are tiny. If you scaled up a grape to the size of Earth, its atoms would be about grape-sized. This means there are a great many atoms in the smallest of things. In our grape, for instance, there are as many atoms as there are stars in the observable universe.

Democritus and his contemporaries had no way of knowing that atoms really existed. There was no physical proof that they did: they were the subject of philosophical musings, an inspired lucky guess. But over the past two centuries, we've got to know them well.

We know, for instance, that the atoms in our grape are encircled by fuzzy clouds of *electrons* – aptly so named because they're electrically charged. These charged clouds give rise to the electrical force that binds atoms to one another and governs their interactions. The branch of science that describes and forecasts these electrical tempests is known as *chemistry*.

But suppose we thought our efforts unfinished. What would happen if we kept on slicing our grape? We'd find that Democritus, whilst brilliant, wasn't quite right. Our atom, far from being 'atomos' – *uncuttable* – could be divided further. We would, perhaps inevitably, split it.

Deep within the electron cloud, we'd find a dense kernel of matter called the *nucleus*. It took us 80 slices to cut our grape down to the size of an atom; it would take another 45 to cut our atom down to the size of a nucleus, because the nucleus is 20,000 times smaller than the atom as a whole.

So, what would we find in the space between the electrons and the nucleus? We wouldn't find air, because air is itself made from atoms. The answer is . . . nothing. Atoms, it turns out, are made mostly from empty space. The world is mostly void. It's a thought that makes my mind explode.

Now, imagine that we continued slicing our grape, just a few more times. What then? We'd find our nucleus is composed of *protons* and *neutrons*. The branch of science that describes these subatomic particles and their interactions is called *nuclear science*. (We could go further and divide our protons and neutrons into quarks . . . but we have to stop somewhere.)

Like electrons, protons are charged. But whilst electrons are negatively charged, protons are positively charged. They say that opposites attract, and whilst that's not necessarily true for humans, it *is* true for charges. Unlike charges attract; like charges repel. The electric forces between negative electrons and positive protons hold the two together.

1 H																	2 He
3 Li	4 Be											5 B	6 C	7 N	8 O	9 F	10 Ne
11 Na	12 Mg											13 Al	14 Si	15 P	16 S	17 Cl	18 Ar
19 K	20 Ca	21 Sc	22 Ti	23 V	24 Cr	25 Mn	26 Fe	27 Co	28 Ni	29 Cu	30 Zn	31 Ga	32 Ge	33 As	34 Se	35 Br	36 Kr
37 Rb	38 Sr	39 Y	40 Zr	41 Nb	42 Mo	43 Tc	44 Ru	45 Rh	46 Pd	47 Ag	48 Cd	49 In	50 Sn	51 Sb	52 Te	53 I	54 Xe
55 Cs	56 Ba	57-71	72 Hf	73 Ta	74 W	75 Re	76 Os	77 Ir	78 Pt	79 Au	80 Hg	81 Tl	82 Pb	83 Bi	84 Po	85 At	86 Rn
87 Fr	88 Ra	89-103	104 Rf	105 Db	106 Sg	107 Bh	108 Hs	109 Mt	110 Ds	111 Rg	112 Cn	113 Nh	114 Fl	115 Mc	116 Lv	117 Ts	118 Og

57 La	58 Ce	59 Pr	60 Nd	61 Pm	62 Sm	63 Eu	64 Gd	65 Tb	66 Dy	67 Ho	68 Er	69 Tm	70 Yb	71 Lu
89 Ac	90 Th	91 Pa	92 U	93 Np	94 Pu	95 Am	96 Cm	97 Bk	98 Cf	99 Es	100 Fm	101 Md	102 No	103 Lr

The periodic table condenses the complexity of the world – grapes, mountains, galaxies – into 118 chemical elements.

Neutrons, as their name suggests, are electrically neutral. They carry no charge.

Slicing our nucleus into pieces would be tiresome work. Whilst *electrical* forces bind electrons to the nucleus, *nuclear* forces glue the nucleus itself together. And those nuclear forces are far stronger than their electrical counterparts. They're mighty enough to overcome the mutual electric repulsion of the protons and keep the nucleus intact. Combined, these atomic forces make the world – mostly void – appear solid.

Atoms come in 118 chemical varieties called *elements*. Each element is defined by the number of protons in its nucleus, and chemists arrange them on the periodic table as such. For instance: hydrogen, element number 1, always has 1 proton in its nucleus, and so occupies the 1st tile; platinum, element number 78, always has 78 protons in its nucleus, and so occupies the 78th tile; and so on.

The Nobel Prize-winning physicist Ernest Rutherford reportedly once quipped that 'all science other than physics is stamp collecting'. Well, if that's true, the periodic table is humanity's finest stamp collection.

On Earth, there are 88 naturally occurring elements.* Our grape was made mostly from half a dozen: hydrogen, carbon, nitrogen, and oxygen, and a dash of potassium, calcium, and a few other metals. Air is a mixture comprised chiefly of three: nitrogen, oxygen, and argon. Water is a concoction of just two: hydrogen and oxygen. Diamond is a crystal of only one: carbon.

The other 30 elements on the periodic table do not exist naturally. These are the *synthetic elements*, which only exist because we, humans, made them by either sticking existing nuclei together or pulling them apart.

Now, as much as I delight in the periodic table, it has a major shortcoming. It only tells half the story, because it fails to consider *neutrons*.

Neutrons have almost the same mass as protons (they're a paltry 0.1 per cent heavier), and by varying the number of them in an atom's nucleus, the mass of the atom changes. We call atoms of the same element with different numbers of neutrons *isotopes*.

The word 'isotope' is another nod to Democritus: it comes from the ancient Greek ἴσος – 'isos' – meaning *same*, and τόπος – 'topos' – meaning *place*. Isotopes occupy the *same place* on the periodic table because they're the *same element*. They differ only in how heavy they are.

There are 118 elements, but there are *thousands* of isotopes. Carbon, for instance, has three: 'carbon-12', 'carbon-13', and 'carbon-14'.† Every atom of carbon, by definition, has 6 protons, but its isotopes have different numbers of neutrons. The number following the element name – the atom's *mass number* – is simply the sum of the protons and the neutrons. Carbon-12 atoms,

---

* The sticklers will remind me that there are actually more like 94 naturally occurring elements on Earth, but I'm ignoring six: technetium (element 43), promethium (element 61), astatine (element 85), francium (element 87), neptunium (element 93), and plutonium (element 94). They exist in such minute quantities that whilst they *technically* exist, they *practically* do not.

† The same sticklers will remind me that there are more than three isotopes of carbon. But I'm ignoring all but three, because they're the only ones that occur naturally on Earth.

therefore, have 6 protons and 6 neutrons; carbon-13 atoms have 6 protons and 7 neutrons; and carbon-14 atoms have 6 protons and 8 neutrons.

Given their electrical neutrality, neutrons have no bearing on the chemical character of an atom. Different isotopes of an element are doppelgängers with identical chemical properties, which is why chemists normally get by without giving them much heed.

But not so in nuclear science. Neutrons are of paramount importance, because they affect the *nuclear* character of an atom. Isotopes of the same element often have different nuclear properties, amongst them the ability to radioactively disintegrate or split in two and release tremendous amounts of energy.

And after pondering these things, we mustn't forget to reassemble our nucleus, put our atoms back together, and rebuild our grape. Then, we'll have earned our wine.

A cloudy week in Paris

MONDAY, 24 FEBRUARY 1896.

Parisian physicist Henri Becquerel is experimenting with the uranium – element number 92 – that he inherited from his father. He's using it to better understand *phosphorescence*, a curious phenomenon where a substance apparently 'charges up' in sunlight and slowly radiates its energy as an eerie afterglow.

Becquerel's experiment is simple: he wraps a photographic plate in thick paper, sprinkles the uranium on top, and basks it in sunshine for a few hours. The thick paper stops the sunlight reaching the plate, but when he develops the photograph, the ghostly image of the uranium appears. The uranium atoms seem to absorb energy from the Sun and re-emit the rays with enough vibrance to shine *through* the opaque paper. How peculiar.[1]

Two days later, Becquerel prepares a second experiment. He wraps another photographic plate in thick paper, sprinkles more uranium on top, and basks it in . . . but no. It's cloudy. With no sunlight to 'charge up' his uranium, he terminates the experiment. Defeated, he stows the photographic paper – uranium and all – in the darkness of his bureau drawer for a sunny day.

The clouds above Paris do not lift for the rest of the week. Thursday . . . grey. Friday . . . still grey. Saturday is no better.

By Sunday, Becquerel's patience is strained. He retrieves the experiment from his drawer and develops the photograph anyway. He expects to see nothing. But there, as clear as if he'd left it out in bright sunshine, he sees the ghostly image of the uranium.[2] How can this be? Where have the rays come from if the uranium had been kept in the dark?

The answer is inescapable: uranium atoms do not get their energy from the Sun. Their energy was there all along. It's *inherent* in the uranium itself, and it has enough vigour to shine through thick paper and fog photographic plates. News of the discovery galvanises the scientific community.

Two years after Becquerel's discovery, German chemist Gerhard Carl Schmidt and Polish-French chemist Marie Curie each independently find another element that does the same thing: thorium (element number 90).[3] Thorium, like uranium, radiates energy into its surroundings without external stimulus. So extraordinary is this phenomenon that no word exists to describe it. Curie coins one: *radioactivity*.

Within two years, Marie Curie and her husband, Pierre, discover two more radioactive elements. They name one (element number 84) *polonium* after Marie's homeland, Poland. The other (element number 88) is so radioactive that it glows in the dark, so they name it *radium* after the Latin *radius*, meaning 'ray of light'. In 1903, they become the first married couple to win a Nobel Prize in Physics, sharing it with Becquerel.

In 1911, Marie Curie would win a second Nobel Prize – this time in Chemistry – for the discovery of polonium and radium. (Pierre died tragically in 1906 after a horse-drawn carriage ran him over, and the Nobel Committee doesn't award prizes posthumously.) She remains the only person to win a Nobel Prize in two separate sciences. Little did she know that they would kickstart a scientific revolution that would turn our understanding of material reality on its head.

Easy as α, β, γ

We know today that radioactivity comes in different varieties. The most common types by far are the ones I encounter the most in my lab: *alpha*, *beta*, and *gamma* radiation, named with another nod to Democritus according to how easily they pass through matter.

Different radioactive elements – or, more precisely, different radioactive *isotopes* – emit distinct types of radiation. All naturally occurring isotopes of uranium, for instance, emit alpha radiation. So do the Curies' polonium and radium. On the other hand, naturally occurring potassium-40 – which comprises 1 in every 10,000 of all potassium atoms on Earth[4] – emits beta and gamma radiation. Grapes, and anything else that contains potassium, such as bananas, potatoes, and building materials, are therefore slightly radioactive.

Atoms that radioactively decay have too many or too few neutrons in their nuclei. They are off-kilter. They radiate energy spontaneously through *radioactive decay* to regain balance. This radiation, like atomic-scale artillery, carries with it enormous energy.

But it turns out that alpha and beta radiation aren't 'rays' at all. They're *particles*, small fragments of atoms.

Alpha particles are hefty and energetic. They're bulky clusters comprising a pair of protons and a pair of neutrons stuck together. This means alpha particles are actually helium nuclei, lacking only the accompanying electrons that would make them fully fledged atoms. They're fired like cannonballs from their parent nuclei at more than 30 *million* miles per hour, and they transmit their energy into their surroundings when they collide with other atoms.

Alpha decay fundamentally changes an atom's nature. The parent atom loses 2 protons and 2 neutrons; this means its mass number decreases by 4. But crucially, by losing the pair of protons, its element number decreases by 2. *This means it becomes a new element.* Radioactive atoms naturally transform from one element into another, without any human intervention. Upon firing out an alpha particle, the parent atom moves 2 spaces down the periodic table.

When uranium (element number 92) alpha decays, it becomes thorium (element number 90); when thorium alpha decays, it becomes radium (element number 88); when radium alpha decays, it becomes radon (element number 86); and so on. The cascade continues down a long chain, from one radioactive isotope to another. Eventually, a non-radioactive isotope is encountered, at which point the cascade stops. In the case of uranium, the chain cascades all the way to element number 82, lead.

The spontaneous transformation of one element to another adds a layer of complexity to working in a nuclear lab. My labmates and I are mindful that our samples change their chemical nature continuously; what we put on the shelf today won't be the same tomorrow.

Since they're so burly, alpha particles don't travel far before coming to a halt. They crash into other atoms and lose their energy quickly. A thin sheet of paper blocks them easily. So does the top layer of your skin. Even a finger's width of air is enough to stop an intense stream of alpha particles dead in its tracks.

Beta particles, on the other hand, are zippy and weigh very little. They're not clusters of protons and neutrons but nimble *electrons*, more bullet-like than their cannonball cousins. They approach light speed, dissipating their energy far and wide as they jostle past the atoms in their surroundings. Beta particles can traverse several metres of air, and a thick layer of plastic or a thin sheet of metal is required to stop them.

Like alpha decay, beta decay changes the parent atom's nature. One of its neutrons turns into a proton; this means the atom's mass number stays the same, but its element number increases by 1. Thus, as with alpha decay, it becomes a new element. Upon firing out a beta particle, the parent atom moves 1 space *up* the periodic table. When radioactive potassium (element number 19) beta decays, it becomes calcium (element number 20); when radioactive carbon (element number 6) beta decays, it becomes nitrogen (element number 7); and so on.*

---

* There's a rare type of beta decay called beta-plus decay whereby a proton turns into a neutron. This releases a positron (the anti-matter counterpart to the electron) and moves the element one space *down* the periodic table.

Gamma radiation is the oddity of the trio. It *is* made from rays, not particles. Gamma rays are individual flashes of light, albeit light that lies so far along the electromagnetic spectrum that it's invisible to our eyes. They travel at light speed and carry tremendous energy, and they're only stopped only by a head-on collision with the nucleus of an atom. As atoms are mostly void, gamma rays shine through matter with ease. To block them, you need something like a thick wall of lead or dense concrete. Sheldon Allman got it right in his 1960 sci-folk number 'Radioactive Mama': *Well, your kisses do things to me in oh so many ways, I feel them going through me, all those gamma gamma rays.* Gamma decay does not change the parent atom's nature. Rather, the atom simply rids itself of excess energy.

Occasionally, I switch on my radiation probe and it erupts into a chorus of *click click clicks* because of beta and gamma radiation coming from a sample a few feet away. One time, a sample *on the other side of the lab* sent my radiation probe into a frantic clatter. I raised an eyebrow at my labmate. 'Yeah, that's a spicy one', he said. At the levels in our labs, though, the exposure is harmless. Besides, we do a good job of blocking beta particles with sheets of aluminium. And when it comes to gamma rays, they're no match for the lead bricks we stack around the samples.

Alchemy

Anybody who has attempted to cut a single atom from a grape will have quickly learned that you cannot get inside atoms using ordinary means, because ordinary things – like knives – are themselves made from atoms. Extraordinary insight requires *extraordinary* tools: radiation.

Kiwi physicist Sir Ernest Rutherford is remembered by history as 'the father of nuclear physics', even though, ironically, he won the 1908 Nobel Prize in Chemistry. He discovered the atomic nucleus, indulged in a spot of stamp collecting by adding the 86th tile to the periodic table (radon), and classified Becquerel's radioactivity into alpha particles, beta particles, and gamma rays. He also derived the mathematical underpinnings of radioactive decay.

In 1919, Rutherford began shelling nitrogen atoms with alpha particles. He observed something bizarre. Occasionally, an alpha particle would smash into the nucleus of a nitrogen atom, and when it did, it broke the nuclear forces gluing the nitrogen nucleus together. Small pieces were blasted away by the collision. Rutherford named his atomic fragments *protons*, thus adding the proton to his already impressive repertoire of discoveries.[5] By smashing protons out of nitrogen nuclei, Rutherford hadn't quite *split* the atom, but he had knocked a chip off the old block.

But what became of his alpha particles after their head-on collision? It took British experimental physicist Patrick Blackett 6 years to uncover the answer.

In 1925, Blackett followed Rutherford in firing alpha particles at nitrogen atoms. (He would win the 1948 Nobel Prize in Physics.) Blackett found that his alpha particles didn't just ricochet off the nuclei after the collisions. They *merged* with them.[6]

Nitrogen is element number 7. Merging it with an alpha particle adds 2 more protons; losing a proton in the blast takes 1 away. That makes 7, plus 2, minus 1. It wasn't element number 7 anymore. It was element number 8, oxygen.

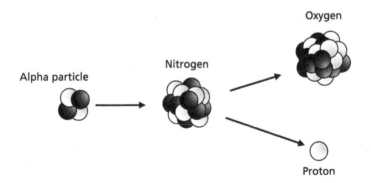

By transforming nitrogen into oxygen, Rutherford and Blackett had changed the fundamental nature of an atom at will. They had sparked the first human-made nuclear reactions, achieving the seemingly impossible and fulfilling the alchemist's dream of turning one element into another, albeit they hadn't *quite* transformed dull lead into glittering gold.

That's what makes nuclear reactions profoundly different to chemical reactions. In chemical reactions – burning methane gas, for example – the atoms are simply reshuffled into new combinations; the carbon and hydrogen in methane (CH_4) react with oxygen (O_2) in air to make carbon dioxide (CO_2) and water (H_2O). The atoms before are the same as the atoms after. They're in a different order, but they themselves don't change.

In nuclear reactions, though, the atoms aren't merely reshuffled. *The atoms change.*

Atom splitting

Firing alpha particles at atomic nuclei is tricky business. They're both positively charged, and their mutual repulsion intensifies as the alpha particle approaches. The electrical forcefield surrounding the nucleus deflects alpha particles off-course. Most swerve and miss entirely, and the repulsion is so strong that some do a complete 180 and hurtle back whence they came. Only occasionally does one strike the dense nucleic kernel head-on and spark a nuclear reaction.

But there's an easier way to get inside atoms: by firing neutrons at them. In their neutrality, neutrons sail towards charged nuclei unperturbed by electrical forces. This is exactly what Italian physicist Enrico Fermi started doing in 1934. Fermi and his colleagues at the University of Rome fired neutrons at atoms in the hope of making new elements, but their target wasn't nitrogen. It was uranium, right at the other end of the periodic table.

Uranium naturally radiates alpha particles. After Fermi bombarded his uranium with neutrons, he found that it fired out beta radiation, too. This meant that his sample was no longer *just* uranium, proving that he'd sparked a nuclear reaction and created a new isotope. But which element was it? Painstaking tests failed to pin it down.

No other element on the periodic table in the vicinity of uranium matched the chemistry he saw.[7]

Moreover, the properties of the beta radiation indicated the presence of more than one new isotope. Fermi and his colleagues hadn't created a new isotope; they'd created new *isotopes*, plural. They were dumbfounded.

Four years later, German chemists Otto Hahn and Fritz Strassmann repeated Fermi's experiments at the University of Berlin. They, too, found a medley of radioactive isotopes in their uranium after they'd bombarded it with neutrons.

They identified one as barium. *Barium*! Uranium is element number 92; barium is element number 56. That's a difference of 36 protons! At the time, it was thought impossible to knock more than a couple at a time out of an atom's nucleus because the nuclear forces glueing them together are so strong. How had they managed to knock out 36? It made no sense. Their disbelief is stark in the paper they submitted to *Naturwissenschaften* on 22 December 1938:

> ... we cannot bring ourselves yet to take such a drastic step which goes against all previous experience in nuclear physics. There could perhaps be a series of unusual coincidences which has given us false indications.[8]

At a loss as to how they'd made barium, they sought help from their friend and colleague, Lise Meitner.

Meitner became professor of physics at the University of Berlin in 1926. She fired neutrons at uranium alongside Hahn and Strassmann until Adolf Hitler's troops poured into Austria and annexed it in 1938. Meitner, who was Austrian-born with Jewish ancestry, lost the protection of her Austrian citizenship. With the help of Hahn in July 1938, she sneaked over the German border into the Netherlands as a refugee.[9]

Meitner and her physicist nephew, Otto Frisch, spent the Christmas period of 1938 fixated on the seemingly impossible barium. By the time Hahn and Strassmann's paper was printed on 6 January, Meitner and Frisch had solved the problem of the impossible barium. The answer hit them like an apple falling from a tree.

Just as opposite charges attract, like charges repel. The electrical forces of repulsion between the protons in an atom's nucleus relentlessly try to blow it to pieces. But attractive nuclear forces are far stronger than electrical forces. They triumph over the protests of the mutually repulsive protons, thereby holding the nucleus together.

However, nuclear forces only work over short distances. You can think of them like glue: the protons and neutrons have to practically touch for them to stick together. Electrical forces, on the other hand, work over large distances: you can think of them as magnets that push and pull, despite being separated spatially. Meitner and Frisch visualised the uranium nucleus not as a solid sphere, but as a liquid drop. The nuclear droplet, perturbed by a striking neutron, starts to wobble.

In one of its fluctuations, the nucleus becomes elongated. This marks the point of no return: the two ends of the elongated droplet are too far apart for nuclear forces to pull them back together; repulsive electrical forces get the upper hand and push them even further apart. The protons and neutrons at each end of the stretched-out nuclear droplet, however, are pulled together into round bulbs by attractive nuclear forces. The nucleus begins to resemble a dumbbell. The two bulbs repel each other further still. The thread joining them becomes thinner, and thinner, and thinner, and then . . . *snap*.

The uranium nucleus splits in two.

A pair of smaller nuclei appears in place of uranium. The fragments, each carrying a share of the uranium's protons, occupy entirely different parts of the periodic table. The uranium doesn't split in half, exactly; it's an asymmetric split. One of them might be barium (atomic number 56), for example, in which case the other would be krypton (atomic number 36).* The barium-krypton duo is just one of many possible pairs.

Hahn and Strassmann used neutrons to split the atom. Meitner

---

* The element numbers of the fragments must add up to the element number of uranium (92).

and Frisch worked out how, and for the first time used a name to describe it: *nuclear fission*.[10]

A uranium nucleus splits – or fissions – into two smaller nuclei after being hit by a neutron.

$E = mc^2$

Recalling the moment he and Meitner had their epiphany, Frisch said, 'Then Lise Meitner was pursuing a separate trend of thought, and was saying that if you really do form two such fragments, they would be pushed apart with great energy.'[11] Meitner was right. After a nucleus splits, the pair of fission fragments are too far apart to feel the pull of attractive nuclear forces, but they're plenty close enough to feel the mutual push of repulsive electrical forces. Propelled by their like charges, the two rocket away from each other at some 8,000 miles per second. It would take about a minute to fly to the Moon and back at that speed.

For her contributions to our early understanding of fission, element number 109 – *meitnerium* – was named in Meitner's honour. But where does this energy come from? Albert Einstein – whose name was bestowed on element number 99, *einsteinium* – explains.

Imagine for a moment a vintage set of weighing scales. With impossibly steady hands, we place a single atom of uranium on one of the pans. On the other pan, we place a pair of fission fragments.

With bated breath, we watch as the balance tipped slowly in

favour of the uranium. A uranium atom, it turns out, is heavier than the fragments it splits into. It *loses mass* when it fissions.

But where does this missing mass go? Einstein's famous equation – $E = mc^2$ – tells us that mass (m, in kilograms) can be converted into energy (E, in joules) and vice versa. The exchange rate, c^2, is the square of the speed of light (in metres per second). Our missing mass wouldn't be 'missing' at all. It would just have been converted into energy.

Upon scrutinising our vintage scales, we'd find the difference in mass would be something like 0.000000000000000000000000004 kilograms (27 zeroes). That's not a lot of mass. But in $E = mc^2$, we get an enormous amount of E from a tiny amount of m, because the numerical value of our c^2 exchange rate is absolutely enormous: 89,875,517,873,681,764.

Plugging our m and c^2 into Einstein's equation yields an E of 0.0000000003 joules (ten zeroes). It's easy to lose yourself in absurd numbers of zeroes, and so we express this burst of energy in units of *electron-volts*, in which case it conveniently becomes 200 million.[12] That's not a lot of energy *per se* – it's 3,000 times less energy than is carried by a single mosquito in flight – but it *is* an astronomical amount of energy for a single atom. And there are 2,500,000,000,000,000,000,000 atoms in a single *gram* of uranium.

By pale comparison, chemical reactions – like burning coal and gas – typically release a couple of electron-volts per atom.

There's as much nuclear energy in a gram of uranium as there is chemical energy in more than a *tonne* of coal. Put another way: if you powered a typical lightbulb using 1 gram of coal, you'd have less than 15 minutes of light; if you made full use of 1 gram of uranium, you'd have light for 30 *years*.[13] Nuclear fuel, when compared to chemical fuel, is in a class of its own.

Chain reactions

Fission fragments aren't the only things blasted away when uranium atoms split in two. Neutrons, too, usually in pairs or triplets, are expelled. And in the presence of other uranium atoms, loose neutrons are like flint-sparks in a haystack.

When a neutron spark collides with another uranium atom, it ignites another fission reaction. This generates more neutron sparks. Then *those* neutron sparks ignite fission in *other* uranium atoms, thus creating *more neutron sparks*. This loop of positive feedback gives rise to a fission chain reaction that, once caught, sustains itself. One fission quickly becomes many.

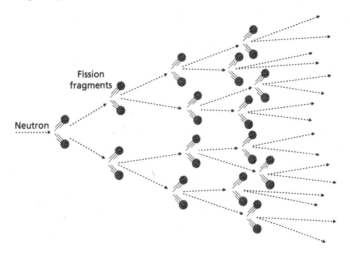

An important number used to describe this neutron propagation is the k-value: the average number of neutrons from a fission that will spark another. The numerical value of k dictates whether a chain reaction will go out with a whimper or off with a bang, or burn along steadily.

Some neutron sparks are absorbed by other atoms. Some miss other uranium atoms entirely. If enough of them meet either of these fates, then k is less than 1, and our chain reaction is *sub-critical*.

Chain reactions where k is greater than 1 are governed by the mathematics of exponential growth. Each fission will spark more than one subsequent split. Tiny numbers of fissions become many in a small number of links, and our chain reaction is *super-critical*.

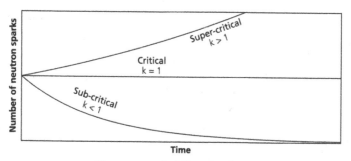

Neutron curves for different k-values.

To illustrate the power of exponential growth, say k equals 2 (the number of links doubles every time). The number of fissions escalates quickly out of control: 1, 2, 4, 8, 16, 32, 64, 128, 256, 512, 1024, 2056...

After 20 links, there are more than a million fissions; after 40 links, there are more than a trillion. After 266 links, you'd run out of atoms in the observable universe. What begins as a single neutron spark can surge into a firestorm.

If our k-value is equal to 1, then our chain reaction is in the Goldilocks zone. Each fission sparks one other on average. In this steady state, our chain reaction is *critical*.

The A-bomb

It dawned on physicists that if they sparked a super-critical nuclear chain reaction, $E = mc^2$ and the mathematics of exponential growth would liberate vast amounts of energy rapidly. Devices that liberate energy rapidly are known as 'bombs'.

A bomb that derived its explosive power from nuclear fission would be a new type of weapon. It would be an *atom bomb*. Such a device had never been seen in the arsenal of any nation. Conventional bombs, which derive their energy from chemical reactions, would detonate with mere whispers by comparison.

With the eve of World War II darkening the skies of Europe, the thought frightened physicists. In the summer of 1939, a letter appeared on United States President Franklin Roosevelt's desk, penned by Einstein and fellow physicist Leo Szilárd, with a grim warning:

> ... it may become possible to set up a nuclear chain reaction in a large mass of uranium, by which vast amounts of power and large quantities of new radium-like elements would be generated. [14]

It wasn't a huge conceptual leap from the idea of chain reaction to the atom bomb:

> ... and it is conceivable ... that extremely powerful bombs of a new type may thus be constructed.

It was possible in principle. Why not in practice? Alarmed, Roosevelt set up a committee tasked with studying the nuclear characteristics of uranium. If such a device could be constructed, it was vital that Nazi Germany didn't build one first. This committee, in time, morphed into the Manhattan Project.[15]

But in 1939, only tiny numbers of atoms had been split in carefully controlled benchtop chemistry experiments. Scaling up the experiment to spark full-blown chain reactions where k exceeded 1 was purely hypothetical. This is where Italian physicist Enrico Fermi re-enters our story.

Super-criticality

Fermi was awarded the Nobel Prize in Physics in 1938 for his work on bombarding uranium with neutrons. Fermi attended the award ceremony in Stockholm in mid-December and used the trip as his chance to flee Benito Mussolini's increasingly anti-Semitic and Nazi-allied Italy. With his family, he took refuge in the USA, arriving mere weeks before the discovery of fission broke in the scientific press.[16] At his new posting at the University of Chicago, Fermi would lead the effort to spark nuclear chain reactions in the world's first nuclear reactor.

But Fermi had a problem: neutrons are born fast. When uranium nuclei burst apart, the neutrons stream outwards at 12,000 miles per *second*, fast enough to circumnavigate the Earth in 2 seconds.[17] At such breakneck speeds, they tend to sail right past their targets. They're like bullets whizzing past a scattering of fridge magnets, not feeling the slightest pull. Only a tiny fraction end up engaging in nuclear fission, and the chain reaction fizzles towards sub-criticality.

To increase the chance a neutron will collide and interact with a nucleus, it must be slowed down, or *moderated*. A slightly slower neutron can be drawn in by nuclear forces. This puts us in the counterintuitive situation where the slower a neutron travels, the more likely it is to cause fission. Slow neutrons yield chain reactions with more vigour.

To slow down his speeding neutrons, Fermi passed them through the mineral form of carbon: graphite. It was cheap, it was easy to get hold of, and it could be purified in huge quantities.

In the end, Fermi's reactor comprised 350 tonnes of graphite and 36 tonnes of uranium. More graphite meant more slow neutrons; more uranium meant more targets for those neutrons. The only place big enough for the contraption was the old squash court beneath the spectator stands of the university's football pitch.[18]

Construction of the reactor began on 6 November 1942. Fermi and his fellow scientists laboured around the clock. And it was gruelling work. Each slippery graphite block weighed 10 kilograms, and there were thousands of them. They were hand-cut and heaved, one on top of the other, into the intricate lattice prescribed by Fermi. The most efficient overall shape of the final structure was determined to be a sphere, and the scientists approximated it as best they could with their large, angular building blocks.

The reactor went in alternating layers: a layer of just graphite, followed by two layers of graphite interspersed with slugs of uranium. Fermi calculated the reactor would go super-critical once the 57th layer had been laid.[19]

As the reactor grew, so too did the wooden scaffold that supported it. The whole thing looked – and I don't mean this disparagingly – homemade. But in a sense, it *was* homemade. It was brand-new technology constructed using nascent knowledge. They were writing the manual as they went.

There was also safety to consider. What if the chain reaction escalated out of control? Fermi, as always, had a plan.

Elements that absorb neutrons without fissioning are called *neutron poisons*. They eliminate neutrons from the chain reaction, thus severing links and depressing k. That's undesirable when you want your reactor to go critical; nuclear scientists to this day work hard to eliminate neutron poisons from their reactor materials. But it's a good thing if you want to intentionally absorb neutrons. If poisons are slid into a reactor, k plummets and the reactor is *scrammed*.

Fermi used cadmium as his neutron poison. It wasn't engineered into the precision rods you'd find in a modern nuclear reactor, though. It was nailed, in sheets, to 12-foot-long planks of wood. The poison planks were pushed by hand into slots running horizontally through gaps in the graphite blocks to control the flow of neutrons. It was rustic . . . but it worked.

There were also slots running vertically through the reactor. Poison planks, suspended by rope, would automatically drop should the neutron flux climb too high.

As a back-up to the automatic scram, physicist Norman Hilberry was given an axe. He later recalled, 'I was ushered to the balcony rail, handed a well sharpened fireman's axe, and told that was it, "*if the safety rods fail to operate, cut that rope.*" '[20]

And as a back-up to the axe-man-back-up, a three-person 'Suicide Squad' was assembled. It was their job to stand on top of the reactor with buckets of cadmium solution, ready to douse the reactor and snuff it out should all other scram mechanisms fail. The leader of the squad, physicist Samuel Allison, later insisted that it was his brawn, not his brain, that qualified him for this ultimate last-ditch failsafe.[21]

On the chilly night of 1 December, the 57th layer of graphite was laid. Fermi stood before his masterpiece. The blocky graphite spheroid – radiant black, 6 metres pole to pole and almost 8 metres wide at the equator – loomed over him.

Chapter 1: Wresting Fire from the Gods 27

2 December 1942. Fermi and his colleagues watch Chicago Pile-1 – the world's first nuclear reactor – go critical at the inauguration of the Atomic Age. You can see the Suicide Squad atop the reactor. Image courtesy of U.S. National Archives and Records Administration/Science Photo Library.

The next morning, Fermi and 48 of his colleagues gathered on the squash court balcony.[22] This was it. A journey that began 46 years previously when Becquerel discovered radioactivity beneath cloudy Parisian skies was about to culminate in the dawn of a new age.

Fermi ordered the host of electrically operated poison planks to be withdrawn from the reactor. The *click, click, clicking* of the neutron counters became a little twitchier. But k was less than 1. It wasn't critical.

Then Fermi ordered that the penultimate poison plank – nicknamed 'Zip' – be retracted. Hilberry stood by the rope that hoisted it out, axe in hand, just in case. The neutron counters began *clicking* as a swarm. The only thing standing in the way of criticality was the final plank. Femi directed 35-year-old physicist George Weil to slide it out of the reactor, inch by inch.

'Pull it to 13 feet, George.'

For 5 strained hours, Weil withdrew the final plank further and further. Each time, k inched towards 1. The *clicking* became more frenzied. Fermi's eyes darted from one dial to another, his brain whirring as his fingers worked his ivory slide rule. With each nudge of the plank, he checked the neutron flux against his calculations. Everything was going exactly as he'd predicted.

'Move it 6 inches', Fermi called from the balcony. Each time, Weil obliged.

And then, the final command. 'Pull it out another foot'.

The reactor flickered to life. The feverish *clicking* of the neutrons blurred into a *cccccccc*. The chain reaction was self-sustaining; k equals 1.0006. Super-criticality. 'The pile has gone critical,' Fermi announced through a grin. It was 15:49 on Wednesday, 2 December, 1942.[23] Dawn broke over the Atomic Age.

The enormous reactor was generating half a watt of power, barely enough to flicker a lightbulb. No fuses blew, no wires sparked – it was unspectacular. But k exceeded 1. Left uncontrolled, things *would* become spectacular, in all the wrong ways. After 90 minutes, the reactor would climb to 1 *billion* watts, killing everybody in the room and melting the pile through the squash court floor.

But Fermi shut it off after 4 minutes. 'Okay – Zip in!'. The poison plank severed the chain reaction. Scram. It was all over. (Neither axeman nor Suicide Squad was needed.)

The first controlled release of nuclear energy was an epoch-defining moment for humanity. Fermi and his colleagues marked the occasion by quietly drinking Chianti from paper cups. They had earned their wine.

※ ※ ※

Prometheus wrested fire from the Olympian Gods and gave it to humankind. With it, we swiftly learned how to kindle brilliance and ignite terror. We use fire to shine light into darkness, abate the cold, and accrue material wealth. We also use it to burn books and raze buildings. The choice of what to do with fire always was, and will forever be, ours.

The moment Fermi's reactor went critical, we wielded a new type of fire. As the fire of old before it, the flame of nuclear fission brought us to the forked road of promise and peril.

One path is paved with energy abundance and leads to a long

future of human flourishing. Nuclear energy, in its potency and profusion, could power our world peacefully and perpetually with little environmental desecration. The other path, cast in the shadows of mushroom clouds, leads to oblivion. When the reactor in Chicago went critical, Szilárd shook hands with Fermi and, tormented by thoughts of the atom bomb, lamented, 'This day will go down as a black day in the history of mankind.'[24]

I think Szilárd was unduly pessimistic.

Humans are curious and inventive. Seeing what things do is part of our nature. Sustaining chain reactions that release vast amounts of energy just happens to be something that certain types of atoms do. Our discovery of nuclear fission was as inevitable as our discovery of fire. It just so happened that nuclear physics reached the brink of Promethean knowledge as war broke out in Europe.

Thus, *atom bombs* and *nuclear reactors* became entwined in the public consciousness. But whilst they're built on the same fundamental physics, atom bombs and nuclear reactors are not the same thing; comparing the two is like comparing a high explosive with a tealight candle. Bombs release as much energy as possible as quickly as possible in uncontrolled chain reactions. Nuclear reactors do the opposite: they release energy slowly, in carefully orchestrated chain reactions, to generate the energy on which peaceful civilisation depends.

Chapter 2. How Nuclear Power Works

It's a hell of a way to boil water.

– Karl Grossman, 1980

Our relationship with energy is a defining feature of our species. No other animal has become acquainted with so many of its guises or acquired need for it in such vast quantities. Nor has any other species uncovered the physical laws that govern how it flows from place to place, making the world whirr as it goes. The long arc of human history is characterised in part by the ways we harvest energy and how we use it.

There was a time when, fuelled only by food, we relied solely on the power of our own bodies to get by. That all changed some 1 million years ago when our pre-human ancestors untapped a new source: the chemical energy sorted in wood, extracted from dead trees by kindling them to flame.[1] Our forebears used this energy to make their lives a little better. They cooked food, told stories by dancing firelight, and kept the cold at bay.

When we traded the hunter-gather lifestyle for farming some 10,000 years ago, our demand for energy went up a notch. Our blistered hands and aching bodies could not meet the demands of agriculture, and so we began extracting energy from beasts of burden. We enlisted oxen to tow our ploughs, donkeys to turn our millstones, and horses to haul our carriages.

In time, we learned to draw energy from new places. We added waterwheels to our repertoire during the first century AD, and we began harnessing wind to turn our millstones in the tenth.[2] We started using fire in new ways, too, such as for baking ceramics from clay and smelting metal from ore.

Wood, animals, water, and wind remained our primary sources of energy for almost a thousand years. During that time, we sustained but a small global population tormented by low life expectancies, devastating infant mortality rates, and hand-to-mouth living. But everything changed when we discovered the potency of fossil fuels: coal, oil, and gas.

In the eighteenth century, the United Kingdom set a new precedent. It transitioned from an economy sustained by agriculture to one forged by industry. Machines replaced hands in the making of things. Trains and barges replaced horses in the hauling of goods. Engines replaced water and wind in the turning of wheels. Production lines replaced cottage industries, and people left the field in favour of the factory. Coal enabled this transformation.

Using fossil energy, society became more productive and living standards were raised. Goods became cheaper, labour went further, and people became wealthier.

By the turn of the nineteenth century, we were piping gas into urban areas. Our invention of the internal combustion engine and the availability of affordable cars required copious amounts of oil, sending our already high energy demands into the stratosphere.

Other nations transitioned from agrarian societies to industrial powerhouses, too, first in the West and increasingly in the rest of the world. Today, we consume 3,100 per cent more energy than we did at the beginning of the nineteenth century. We consume more and more each year.[3]

By learning to harvest massive amounts of energy, we inaugurated the modern world: an increasingly technological civilisation – 8 billion people strong – with a rapidly improving quality of life practically everywhere. In the past century, we hauled billions of people out of extreme poverty, more than doubled the average human lifespan, and reduced child mortality from more than 1 in 3 to less than 1 in 23. Since the dawn of the new millennium, over 2 billion people have gained access to clean drinking water, 4.3 billion have connected to the internet, and the average person has become 50 per cent wealthier. Never in human history has there been such a rapid betterment of living standards as in our present epoch. There have been blips and bumps along the way – and those gains aren't felt everywhere all

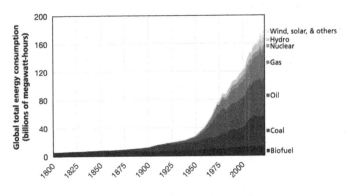

Global energy consumption by source.[4]

the time — but the clear direction humanity as a whole is heading in is 'up'.[5]

And none of it would have been possible without energy abundance. Energy use and national wealth are tightly correlated. Developed nations with low energy needs do not exist. Human progress and energy consumption are lines drawn in parallel:[6]

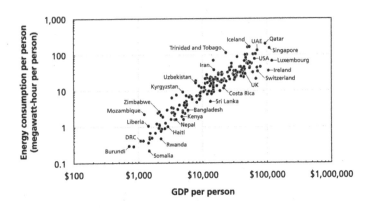

Making as much energy as possible available to as many people as possible ought to be our collective goal. In serving that goal, we must also inflict the minimum harm necessary on human health and the natural environment. Striving to meet these ideals is necessary and ethical, and represents one of the biggest challenges of the twenty-first century.

A friend with many faces

Of energy's various forms – heat, light, motion, sound – there is one that stands in distinction: electricity. Its versatility is unmatched. It conditions the air temperature of hospitals, homes, and offices. It chills food in fridges and roasts it in ovens. It illuminates rooms, animates machines, and powers production lines. It turns the wheels of cars and buses, and conjures music from speakers and headphones. It beams information through our global communications systems and transmits our voices through telephones.

More people plug into the grid all the time. A quarter of a century ago, fewer than 4.5 billion people enjoyed access to electricity; that number has since risen to 7.3 billion, encompassing 91 per cent of the world's population. Hannah Ritchie, Pablo Rosado, and Max Roser from *Our World in Data* point out that every day since the dawn of the new millennium, more than 300,000 people on average gained access to electricity for the first time. Every. Single. Day.[7]

The world's annual electricity generation has tripled since the mid-1980s. Today, we generate nearly three-quarters of it in power stations by a straightforward process: we draw heat from fuel, the heat boils water to steam, and the steam drives an electricity turbine.[8] We have two choices as to how we kindle that heat: extracting chemical energy from *fossil* fuels or nuclear energy from *fissile* fuels.

Fossil fuels dominate the global electricity supply. As a share of total electricity generation, they've never dropped below 60 per cent. Whilst this electricity helps economies flourish, it comes at a terrible health and environmental cost: the emissions released as a by-product of burning fossil fuels cause the climate to change, the oceans to acidify, and the air to become polluted.[9]

As an alternative to setting fossil fuels on fire, we can draw heat from fissile fuels by sending them critical. In stark contrast to their fossiliferous counterparts, *fissile fuels don't emit air pollution*. Climate-warming and ocean-acidifying gases such as carbon dioxide – and lung-blackening soot – literally do not enter the nuclear equation. This makes nuclear energy a fount of the emissions-free electricity that we so desperately need.

Atoms for Peace

Nuclear science advanced quickly in the early decades of the Atomic Age. Within four years of fission's discovery, Fermi sent his reactor critical; just nine years later, in December 1951, scientists in Idaho used a nuclear reactor to illuminate a string of four lightbulbs.[10] If it was possible to power lightbulbs using uranium, why not the world?

The optimism during those early years was palpable. In what became one of the most consequential orations of the twentieth century, United States President Dwight Eisenhower pronounced before the General Assembly of the United Nations in 1953:

> ... peaceful power from atomic energy is no dream of the future. That capability, already proved, is here – now – today.[11]

'Atoms for Peace' – with its grand ideal of using nuclear science to serve the betterment of humanity – mesmerised policymakers and energy leaders. This ideal persists amongst nuclear scientists. Eisenhower's speech precipitated the establishment of the International Atomic Energy Agency, of which almost every nation on Earth is a member today.[12] The agency, which promotes the safe and peaceful use of nuclear technology, outlines its purpose in the second paragraph of its statute:

> The [International Atomic Energy] Agency shall seek to accelerate and enlarge the contribution of atomic energy to peace, health and prosperity throughout the world.[13]

Three years after Eisenhower's speech, Queen Elizabeth II – then only four years into her reign – stepped off the Royal Train onto a red-carpeted platform at Sellafield railway station in northern England. It's the same station I pull into when I catch the train to work. She wore a blue velvet coat to brace herself against the characteristic Cumbrian chill. The station walls were festooned with flowers. A chorus of welcomes followed the royal procession as it drove across Sellafield towards Calder Hall Nuclear Power Station. On that day, Sellafield was the most important place on the planet.[14]

'Today . . . all of us here know we are present at the making of history,' she announced in her clipped accent. The hushed crowds held on to her every word.

A short distance from her podium lay the core of Calder Hall's reactor, inside which billions of uranium atoms were splitting apart every second in a precisely orchestrated chain reaction. Like coiled springs let loose, their atomic nuclei were bursting apart and liberating the energy pent inside.

Calder Hall's primary purpose was initially to support Britain's atomic weapons programme. But the engineers at Sellafield decided to try something new and domestic alongside: they transformed the surplus nuclear energy – which otherwise would have dissipated into the air as heat – into electricity.

'It is with pride that I now open Calder Hall, Britain's first atomic power station.' With that, Queen Elizabeth II pulled a lever, diverting nuclear electricity into Britain's National Grid for the first time. Electricity flowed 15 miles up the coast to Workington, which became one of the first towns in the world to be nuclear-powered. Nuclear electricity spun washing machines and turned vinyl records.

Whilst other nuclear reactors existed at the time, Calder Hall was the first to operate commercially.* By the end of the 1950s, a fleet of nuclear reactors, 11-strong and spanning four nations – the UK, the USA, France, and Russia – was powering towns and cities.[15] It was a sign of things to come.

---

* Two years earlier, the Obninsk nuclear reactor in the Soviet Union had become the first in the world to deliver energy into a grid. It was 40 times less powerful than Calder Hall, though, and didn't operate commercially. Calder Hall is considered widely as the first 'full-scale' nuclear power station. Claims to be the 'first' in anything are often messy and debatable.

Fissile fuels

Nuclear power stations turn heat into electricity. They generate their heat inside nuclear reactors by splitting atoms in the *fuel*, controlling chain reactions with neutron *poisons*, and slowing down neutron sparks with *moderators*.

Uranium is the fuel (with rare exceptions, which we'll get to in Chapter 7), and it normally assumes its simplest oxide form, uranium dioxide (UO_2), a jet-black powder that resembles coal dust. The uranic powder is pressed into grape-sized pieces and sintered into hard ceramic pellets. From one pellet, a few dozen of which would fit easily in the palm of your hand, a typical nuclear reactor can generate as much electricity as a tonne of coal. Here's one to scale:

The pellets are stacked in sheaths of corrosion-resistant metal, typically a zirconium alloy, to create a fuel rod. Zirconium is *neutron-transparent*, and thus allows free passage to neutron sparks so they might find other uranium atoms to fission. Each fuel rod – 3 to 4 metres long and as wide as a finger – contains a few hundred pellets.[16]

A typical reactor burns through about *50,000* fuel rods – together housing 30 tonnes of uranium – every year. To generate the same amount of electricity in a coal-fired power station, you'd burn through more than 2.5 million tonnes of coal.[17]

And, just like Fermi, we use control rods made from neutron poisons to fine-tune k. These days, we tend to make rods from boron, as it's one of the most powerful and least expensive neutron poisons on the periodic table, but we still use good old cadmium sometimes. Tweaking the power output of a nuclear reactor is as simple as moving the control rods in and out of the core: in, and the reactor winds down; out, and the reactor revs up. Nuclear reactors generate the electricity we need exactly when we need it.

All nuclear power reactors serve the same end – generating electricity by splitting atoms – but they come in different types. The half a dozen or so varieties of commercial reactors fall into two families, based on their neutron moderator: water or graphite. Just 1 in 20 use graphite today, which makes water-moderated reactors the second-biggest source of emissions-free electricity in the world, after hydroelectric dams.[18]

Hot water

The *pressurised water reactor* is without doubt the most successful type of nuclear reactor. Since its commercialisation on the banks of the Ohio River in Pennsylvania in 1957, it has become the most prolific design in the world. The 300 or so pressurised water reactors spread across 27 nations today represent more than three-quarters of the world's nuclear capacity.[19]

Pressurised water reactors are aptly named, because – you guessed it – they operate under immense pressure. The scalding core is encased in a pressure vessel some 12 metres high and 5 metres across, with walls forged from steel almost a foot thick.[20] The whole thing is full of water. They're like giant pressure cookers.

The piping fuel rods, made hot by the energy from nuclear fission, super-heat the water in the core to 325 °C. That's hot enough to melt lead. But the pressure – all 150 crushing atmospheres of it, just shy of the pressure you'd experience a mile beneath the ocean – prevents the water from boiling.[21] (Water boils at about 340 °C at those pressures.)

But if water can't boil, it can't turn to steam. And if it can't turn to steam, it can't drive a turbine. How, then, does a pressurised water reactor generate its electricity?

The core is in fact part of a circuit, the *primary circuit*. A pipeline conveys the super-heated water from the core through a heat exchanger – basically a giant radiator – and then returns the cooled water to the core to be reheated. This serves a dual purpose: it stops the core from getting too hot, and it passes the heat into the *secondary circuit*. The water in the secondary circuit is at low pressure, allowing it to boil, turn to steam, and drive a turbine.

Finally, a chiller condenses the steam downstream of the turbine, and the liquid water is pumped back into the heat exchanger for another boiling.

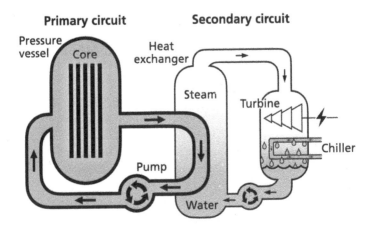

Everything in moderation

Fermi used carbon in the form of graphite to slow down his neutrons. But there's another element that moderates them, too: hydrogen, most commonly found bonded to oxygen atoms in water molecules (H_2O).

One of the smart things about a pressurised water reactor is that the water flowing through its core acts as a coolant *and* a moderator; the water draws heat from the fuel rods *and* moderates the neutrons. It's win-win. This quirky design creates a *negative temperature coefficient*, which makes these reactors incredibly safe.

When k is greater than 1, the rate of fission increases. This causes the core to get hotter and hotter. Left uncontrolled, the nuclear chain reaction would run amok, and the soaring temperature would turn the reactor into a radioactive pipe bomb.

But not in a water-moderated reactor. As the core's temperature rises, the moderating water expands slightly. The atom-sized gaps between the water molecules widen as they push a little further away from one another. This grants easier passage to the streaming neutrons. Fewer collide with the water molecules, which means fewer slow down. The neutrons, on average, become faster. Recall: fast neutrons are unlikely to split uranium atoms. The value of k, therefore, dips below 1, the chain reaction peters out to sub-criticality, and the core cools off. Pipe bomb averted.

This conspires to create the counter-intuitive situation where an increase in temperature slows down the nuclear reaction. It's a negative feedback loop in which the heat cannot soar out of control. And when the reactor cools down, the moderating water contracts, more neutrons are slowed down, and k climbs above 1 to super-criticality. Then, the reactor gets hotter again; water-moderated reactors self-regulate, like they're controlled by a thermostat. It's an engineering stroke of genius.

But there's a problem with using water as a moderator.

Hydrogen exists naturally as two isotopes: hydrogen-1 (so-called 'light hydrogen') and hydrogen-2 ('heavy hydrogen'). Light hydrogen is a worse neutron moderator than its heavier counterpart. In fact, light water – where the H's in H_2O are light hydrogen – isn't a good enough moderator to send natural uranium critical. But heavy water *is*.

The rub with heavy hydrogen is its rarity. If you painstakingly counted 6,500 hydrogen atoms from a bucket of water, you'd find only 1 was heavy. The other 6,499 would be light.[22] Natural water – from the ocean, from taps, from the water in wine – is overwhelmingly *light*.

Pressurised water reactors use natural water as their moderator, which is why they sometimes go by the name 'light-water reactors'. The quenching effect of light water makes it impossible for k to climb above 1. And when k can't climb above 1, chain reactions can't get going. To get around the problem, nuclear engineers have an isotopic trick up their sleeves: *uranium enrichment*.

Enrichment

There are two naturally occurring uranium isotopes: uranium-235 and uranium-238.* Of the two, *only uranium-235 is fissile*. It's the only one that sustains chain reactions using slow neutrons.

As was the case with hydrogen, it's as though nature conspires against us: the more useful isotope is the rarest. If you hadn't run out of patience counting all those hydrogen atoms, you might dig up 1,000 uranium atoms from the ground and count them, too. You'd find that only 7 were fissile uranium-235. The other 993 would be inert uranium-238.[23]

Using natural uranium as fuel is like trying to ignite a bonfire where 99.3 per cent of the sticks won't ever catch. You could still get it going, but it would be difficult. It was one of the reasons Fermi struggled to ignite his reactor; it took 350 tonnes of moderating graphite to overcome the stubbornness of his natural uranium. Using light water as a moderator just adds damp to the fire.

A simple fix – and one employed in most reactors today – is to increase the proportion of uranium-235 in the fuel. This process, called *enrichment*, vastly increases the fuel's potency.

Enriching uranium to the point of super-criticality was one of the most daunting technological hurdles encountered during the Manhattan Project. Uranium-235 and uranium-238 both have 92 protons. They're chemically identical and therefore can't be separated *chemically*. Therefore, the two isotopes must be separated *physically*, by exploiting their slight difference in mass. We do this today using a *gas centrifuge*. As its name implies, a gas centrifuge requires uranium to be in gaseous form. But uranium isn't a gas; it's a metal, all the way up to its blistering boiling point of 4,100 °C. We therefore cheat the system by combining uranium with fluorine to form *uranium*

---

* There are three others (uranium-233, uranium-234, and uranium-236), but I'm ignoring them because they exist in vanishingly small quantities.

hexafluoride – UF$_6$ in chemical shorthand but known affectionately as 'hex'. And hex *is* a gas.[24]

The two isotopes of uranium produce two types of hex: light hex, where the 'U' in **UF$_6$** is uranium-235, and heavy hex, where the 'U' is uranium-238.

The mixture of light and heavy hex is puffed into vertical cylinders, similar in size and shape to the pipes of a church organ. The pipes are spun at 60,000 revolutions per minute.[25] At those speeds, the walls of the cylinders pull about 400,000 g-force. Heavy hex throws its weight around a little more than light hex, and so preferentially squishes itself to the outside wall. The light hex, which weighs three neutrons less, is muscled towards the cylinder's centre. *Voilà*, the isotopes are physically separated.

Light hex is tapped off from the centre of the spinning pipes. What began as hex comprised of uranium-235 in its natural abundance – 0.7 per cent – is now enriched slightly. By running the uranium through the process again, and again, and again, it becomes more, and more, and more enriched.

Vast halls of spinning pipes enrich the uranium in tandem, all the way up to about 5 per cent uranium-235. At that point, the uranium is stripped of its fluorine atoms, combined with oxygen to form uranium dioxide powder, and sintered into fuel pellets. The pellets are loaded into a light-water reactor where they generate electricity. Job done. The carbon intensity of the electricity required to spin the pipes depends on the local composition of the grid; either way, the amount of electricity required is tiny compared to the energy drawn from the finished product.

I work with enriched uranium all the time, albeit normally by the nanogram. The biggest sample by far I've handled amounted to a few crumbs. It was the kind of sample that you'd lose to the four winds should you sneeze on it. Even so, its immense energy-density meant that if I'd pressed and sintered the smidgen into a fuel pellet, I could have generated enough electricity in a light-water reactor to power a lightbulb for five years. From a comparable amount of coal, I'd scrounge enough electricity to power the lightbulb for about 100 seconds.

Now, for every puff of hex that's enriched in a centrifuge, there's a commensurate 'hex tail' that has diminished in uranium-235. These leftovers are known as *depleted uranium*.

Depleted uranium is run back through the centrifuge repeatedly to extract the remaining uranium-235. But harvesting every last atom isn't economical. There comes a point when it isn't worth giving the depleted uranium another whirl (generally at around 0.1 per cent uranium-235), at which point it's normally put in long-term storage. The global stockpile of depleted uranium exceeds 1.2 million tonnes.[26]

Enrichment is a means to an end. It's vital if the quenching effect of light water is to be surmounted. But there's another way to solve the problem. Instead of increasing the fuel's potency by enriching it in uranium-235, we can increase the effectiveness of the moderating water by enriching it in heavy hydrogen. This is exactly the approach taken by the pressurised light-water reactor's big sister: the pressurised *heavy*-water reactor.

Heavy water CANDU it

Pressurised heavy-water reactors work in pretty much the same way as pressurised light-water reactors, but instead use *heavy* water as their moderator.

Whilst more than 99.9 per cent of all hydrogen atoms in natural water are hydrogen-1, heavy water is more than 99 per cent *hydrogen-2*.[27] The difference between light water and heavy water is so stark that if you nursed a glass in each hand, you'd be able to tell which is which from their weight alone. And if you put heavy-water ice cubes your glass of lemonade, they'd sink.

Heavy water moderates neutrons so well that it's enough to send natural uranium super-critical. It even brings *depleted* uranium to life. There's no need for any enrichment. No nation has embraced this type of nuclear reactor more than Canada.

Canada practically invented heavy-water reactors in the 1950s and 60s with its so-called 'CANDU' – '**Can**ada **d**euterium **u**ranium' – reactors. (Deuterium is the fancy name for heavy hydrogen, and the nuclear industry loves a forced acronym.) Canada married its rich indigenous uranium reserves with its expertise in elevating the proportion of hydrogen-2 in its moderator. Today, heavy-water reactors represent every nuclear reactor in Canada's 19-strong fleet. It exported the technology to five other nations, too: India, Argentina, South Korea, China, and Romania.[28]

Electrical yardsticks

The currency of electrical energy is the *watt-hour* (equivalent to 3,600 joules). It takes about 25 watt-hours to boil enough water for a cup of tea. It takes 250 watt-hours to power my desk lamp for a day. And it takes 2,500 – or 2.5 **kilo**watt-hours – to drive 9 miles in a Tesla.[29]

For instance, 1 kilogram of uranium fuel enriched to 3.5 per cent generates 360,000,000 watt-hours or, most simply, 360 **mega**watt-hours.[30] Calder Hall Nuclear Power Station churned out 1,000 megawatt-hours – or 1 **giga**watt-hour (1 billion watt-hours) – of electricity every 4 hours. The total annual electricity consumption of a typical European nation falls into the tens or hundreds of **tera**watt-hours (1 trillion watt-hours) ballpark. There's a cheat sheet for these numerical prefixes in the Appendix, if you find yourself scratching your head over the coming pages.

How much electricity does a person need? It's a tricky number to pin down because different nations use different amounts. An average Norwegian, for instance, uses 26 megawatt-hours per year. An average American uses half that amount. An average citizen of the European Union uses half as much again.[31]

A good yardstick – and one that I'll use throughout this book – is the annual electricity needs of an average person living in an OECD nation. Almost all 38 OECD nations rank 'very high' on the Human Development Index, and they take more than three-quarters of the top 40 spots on the Human Freedom Index.[32] They're geographically disparate and culturally variegated, too: they span from the southern tip of South America to the northern reaches of North America, encompass everywhere in between (and including) Mediterranean and Nordic Europe, and include parts of the Middle East, East Asia and Oceania.

The OECD nations are pleasant places to live. Their high standards of living – and the large amounts of electricity they use to maintain those standards – are what the developing world is heading towards rapidly.

So, how much does it take to electrify an OECD nation? A typical citizen uses 8 megawatt-hours of electricity every year. By comparison, the global average – which, incidentally, is rising year on year – is 3.7 megawatt-hours.[33]

Eight megawatt-hours. That's our yardstick. Now we can put nuclear reactors in context.

The mighty atom

Often, a single nuclear power station hosts several reactors. Calder Hall, for instance, hosted four, which generated enough electricity every year to sate the needs of 200,000 people, before they were switched off in 2003 after 47 years of duty.[34] That's not bad for 1950s technology. But the nuclear reactors we've built since are far more powerful.

In the 1960s, a single light-water reactor typically generated enough electricity for 200,000 people. By the 1970s, enough for over 700,000 people. By the 1980s, almost a *million* people.[35] Even after accounting for downtime, a typical reactor today serves the electrical needs of more than a *million* people. It does this by burning through 80 kilograms of uranium fuel each day.[36] I've burned through more wood than that in a single evening by a campfire. Such is the potency of fissile fuels *versus* chemical fuels.

When we bundle multiple reactors into the same power station, enormous amounts of electricity flow from a single place. In the mid-1970s, a pair of reactors at the Peach Bottom Atomic Power Station in Pennsylvania went live. Over the next decade, they together generated enough electricity for 1.3 million people.

And they got better. Or rather, we got better at running them. In the 1990s, the pair at Peach Bottom generated enough electricity for 1.9 million people; in the mid-noughties, they generated enough for 2.3 million; nowadays, they generate enough for 2.7 million people.[37] That's enough electricity to serve the population of Chicago. Peach Bottom is not unusual in this respect; we often draw more electricity from aged power stations than we did when they were brand-new because the people running them get better at it.

More recently, in 2019, a sixth reactor – Yangjiang-6 – revved up in China's largest nuclear power station in Guangdong province. Together, the sextuplets generate enough electricity for 6.3 million people, setting the standard for what's possible in a modern nuclear power station.[38]

There are more than 400 nuclear reactors spread across 32 nuclear nations today. They span every continent in the world bar Antarctica. But of the 32, the top 5 – the USA, China, France, Russia, and South Korea (in that order) – generate 73 per cent of the world's nuclear electricity. The next 5 – Canada, Japan, Ukraine, Spain, and Sweden (in that order) – generate the next 13 per cent. The distribution of nuclear electricity is so top-heavy that the bottom 27 nuclear nations generate less electricity than the USA alone.[39] As the old saying goes, we harvest the majority of the peas from a minority of the pods.

But when it comes to nuclear's *relative* contribution to the electricity supply, there's no place like Europe. All 10 nations that generate more than a third of their electricity from nuclear are European. And France, Ukraine, and Slovakia use it to generate the majority of their electricity. On a continent-wide scale, Europe generates around a fifth of its electricity from nuclear, making it the biggest source of emissions-free electricity. It's bigger than solar and wind combined.[40]

Sellafield is like a working museum. You'd easily get lost if you didn't know your way around. A network of roads and footpaths, sprawling over a square mile, connects its thousands of buildings. If I ever have time to spare, I meander my way through this nuclear metropolis to one of the seven nuclear reactors dotted about the site. Of the seven (none of which still work), four are bundled into my favourite corner of the entire facility: Calder Hall Nuclear Power Station. I'm always excited to see it, but I walk away tinged by slight sadness.

Calder Hall is past its prime. The cheering crowds and flashing cameras of 1956 are gone. Sea spray and the unforgiving Cumbrian weather have made Calder Hall's once-proud exterior grotty; and its decrepit heat exchangers, which dutifully passed nuclear heat to electricity turbines, stand rusty. Its vast turbine halls lie silent, save for the wind that blows through their broken windows. Its distinctive cooling towers were demolished in 2008. And whereas thousands of nuclear professionals once called it 'work', flocks of Sellafield seagulls now call it 'home'. But the power station still stands as a monument to the power of science to bring technological dreams into reality. Calder Hall ought to be a World Heritage Site.

Following the wiring of Calder Hall into Britain's grid, global nuclear power arrived in three waves. North America dominated the first, Europe the second, and Asia the third. Like all waves, their edges are fuzzy, but they look like this when plotted as a bar chart.

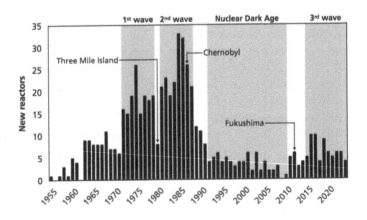

The number of nuclear reactors connected to the grid each year since the dawn of the Atomic Age.[41]

The first wave rolled in during the 1970s and saw 161 reactors revved up within the decade, up from 71 in the 1960s. More than one in three were in North America. The wave peaked in 1974 when, globally, we built 26 in a single year. Whilst we danced to ABBA, we fell in love with nuclear.

Then in 1979, the Three Mile Island accident happened in Pennsylvania. The first wave broke, and the number of new reactors sank to single digits. It damaged the USA's nuclear legacy permanently; it would be 35 years before it built another reactor. But the global swell recovered quickly. By the first year of the 1980s, the number of new reactors was back in the twenties, and the first sign of the second wave was rolling through grids across Europe.

The second wave was bigger than the first. More than half of the reactors were built in Europe, with more than a fifth built in France alone. It peaked at an all-time high in 1984, when we switched on 33 reactors in a single year; 1985 kept the momentum going with 32.

Then, in 1986, Chernobyl happened. Our confidence in the atom broke, and the second wave collapsed. We built more nuclear reactors in the 1980s alone than we've built since.

Thus began the Nuclear Dark Age. In 1990, the world retired more nuclear reactors than it built, submerging them below replacement level for the first time. The number of new reactors we built each year sank steadily through the 1990s and into the new millennium. In 2008, when we revved up a grand total of none. That hadn't happened since 1962.

Then, slowly, the tide began to rise. In 2009, we connected one new reactor to the grid. And then in 2010, another five. And then another six the year after. But then a tsunami, triggered by an earthquake off the Pacific coast of Japan, rocked the boat; Fukushima swallowed the growing wave, and the nuclear resurgence faltered before it reached full height.

Since then, there's been a hint of a third wave, but it was nothing like the two that came before. After Fukushima, nuclear newbuilds numbered six per year on average. China dominates the third wave (52 per cent of new reactors since 2010) with small contributions from Russia (13 per cent) and South Korea (8 per cent).

The third wave is more of a ripple than a swell. Most of the world's nuclear capacity is a legacy of the twentieth century. Calder Hall's weary exterior reflects the current state of nuclear power globally. Neglect crept in. Energy leaders looked away. The world had second thoughts.

Whilst there are powerful new reactors in various stages of assembly, like those in Yangjiang, the global picture is one of decay. Over the 1970s and 1980s, the world built a new nuclear reactor every 19 days on average; over the past 20 years, it built one every 2.5 *months* on average. And between 2002 and 2021, more nuclear reactors retired than opened: 98 up *versus* 105 down.[42]

But in a sense, it's not the number of reactors that matters. It's the amount of electricity they collectively send into the grid. Is replacing scores of old reactors with a smaller fleet of more powerful ones really such a terrible thing?

Since the new millennium, nuclear's share in the global electricity supply has declined steadily from 17 per cent to 9 per cent. It's not that we generate less nuclear electricity now than we did back then – the amount of electricity we generate annually in nuclear power stations has wobbled around the 2,600 terawatt-hour mark for almost a quarter of a century. But during that time, the world's electricity demand almost doubled. Of the sources that rose to meet our increasing demand, chief amongst them are the fossil fuels, which increased by 80 per cent. Wind and solar increased, too, but they still generate five times less electricity than fossil fuels.[43]

And the world's nuclear reactors are ageing quickly. In 1990, the average reactor wasn't even a teenager (it was 11.3 years old). By 2000, it had come of age (18.8 years old). By 2010, it had reached young adulthood (26.3 years old), and today, it knocks on the door of middle-age (almost 32 years old).[44] In the coming decades, more will retire. If we don't replace old reactors with new ones, nuclear's contribution to world energy will wane as our demand for energy continues to grow.

The International Atomic Energy Agency's most optimistic projection shows nuclear's share in world energy will increase by just 1 percentage point by 2050. Realistically, it will probably decline.[45]

But there is strong political and ideological opposition to reversing the decay. Green parties across Europe campaign continuously against new and existing nuclear power stations. A German-led coalition of seven European nations opposes the idea that nuclear-derived hydrogen fuels should be recognised as 'low-emissions', even though they have low carbon footprints, comparable to renewable-derived hydrogen fuels. In 2022, Austria went so far as to take legal action against the European Union when it included nuclear on a list of 'green' investments. Scotland's last nuclear reactor will close in 2028, and it doesn't plan to build any more. For 40 years, the construction of nuclear power stations has been outlawed in Denmark. Australia, too, has laws explicitly prohibiting the construction of nuclear power stations and the fabrication of nuclear fuel. Spain will shut down all its nuclear power stations by 2035. Investors frequently exclude nuclear from their green investment portfolios and major environmental groups campaign against it.[46]

Where vehement opposition is absent, enthusiastic support seldom takes its place. There is an overwhelming feeling of apathy.

Therein lies a colossal contradiction: we're changing Earth's climate because of our utter dependence on fossil fuels, and yet, the world is ambivalent towards harvesting emissions-free energy in nuclear power stations. We're trapped in this situation precisely when we need massive amounts of emissions-free energy the most.

And meanwhile, a growing number of nations are striving to fundamentally rewire how we power our technological world by deploying the biggest mobilisations of talent, resources, and public attention since World War II towards one goal: net zero.

Chapter 3. Unreliables

A 100% renewable energy future IS possible.

– Greenpeace International (@Greenpeace), 2023

In 2019, the British government amended the Climate Change Act so its first paragraph reads:

> It is the duty of the Secretary of State to ensure that the net UK carbon account for the year 2050 is at least 100% lower than the 1990 baseline.[1]

This Act of Parliament legally binds the UK to reduce its net greenhouse gas emissions to nought by the year 2050. Since it became enshrined in law, this policy has become known widely as 'net zero'. The UK was the first major economy in the world to pass such legislation. Others soon followed.

The European Union passed the *European Climate Law* in 2021. Other nations – including Canada, Japan, Luxembourg, Nigeria, New Zealand, Colombia, South Korea, Australia, Switzerland, Kazakhstan, and Chile – have passed laws committing to net zero by around mid-century, too. Not to be outdone, Germany and Sweden have both promised it by 2045. Short of writing it into law, dozens more nations – including India, the USA, and China – have pledged net zero in policy documents or official statements. The British Antarctic Survey has committed to reaching net zero across its entire operations by 2040, taking net-zero aspirations to every continent on Earth.[2]

Net zero is necessary if we're to lighten our species' footprint on the climate. Yet despite the bold ambition, the litany of bromide promises, and the roaming circus of climate change conferences, there's a glaring problem: there isn't a single developed nation that has achieved net zero so far. Not one. And there isn't a single one that has agreed on a coherent plan on how to achieve it.

But one thing is for sure: the only hope we have of reaching net zero is to radically change the ways we harvest energy. Climate change isn't an environmental challenge. It's an energy challenge.

In the first half of the twentieth century, annual global carbon dioxide emissions tripled, from 2 billion tonnes to 6 billion. In the second half, they more than quadrupled to 26 billion tonnes. Today, they exceed 37 billion tonnes. And they're still rising.[3]

But we don't emit this carbon on a whim. By far the largest sources of emissions are the fossil fuels – coal, oil, and gas – that make our world function. They meet more than 80 per cent of the globe's energy demand and are responsible for 95 per cent of carbon emissions.[4] Talk about a double-edged sword.

The only hope any nation has of decarbonising its energy supply and mitigating the effects of climate change, therefore, is to leave fossil fuels in the ground. We must replace them with emissions-free energy sources. The scale of this challenge, compounded by the fear of not rising to meet it, is making our energy leaders flounder.

In what has become a familiar lamentation, United Nations Secretary-General António Guterres prophesied the extinction of humanity in 2018: 'climate change is quite simply an existential threat for most life on the planet – including, and especially, the life of humankind.' Five years later, at the COP28 climate conference in the United Arab Emirates, he proclaimed, 'Earth's vital signs are failing'. These doomsday sentiments are widespread. Sir David Attenborough warned that 'the collapse of our civilisations and the extinction of much of the natural world is on the horizon.' Former US President Joe Biden declared that 'climate change is the existential threat to humanity.'[5]

Transitioning away from fossil fuels ought to propel humanity into a more prosperous, sustainable future; getting it wrong will impoverish people, erode living standards, and destabilise our energy systems whilst failing to solve the problem. To achieve net zero *and* give society the energy that allows it to flourish, we must reject rash policy decisions based on alarmism and magical thinking.

If we want to achieve net zero at all, then we have no choice but to decarbonise our energy supply; if we want to achieve it by 2050, then we must do it quickly; and if we want to achieve it without replacing climate change with another set of problems, then we have to do it smartly. To make a success of net zero in a warming world, cool heads must prevail.

Nowhere near zero

News articles, industry pledges, and, admittedly, the pages in this book thus far tend to focus solely on electricity. But *electricity* and *energy* are not the same thing. Rather, electricity is just one component of our total energy budget. Returning to our electrical yardstick: an average person living in an OECD nation uses 8 megawatt-hours of electricity annually. But their total annual *energy* consumption is 46 megawatt-hours.[6] How is this energy budget sliced and spent?

Global energy consumption, and the downstream emissions, falls into three main categories: buildings, transport, and industry.

Homes and public buildings – offices, schools, libraries, shops, cafés, pubs, banks – together consume almost a third of humanity's energy and are responsible for 27 per cent of its carbon emissions. Just under two-thirds of those emissions come from people's homes.[7]

Modern buildings require a constant flow of electricity, and we generate most of it – globally and within the OECD nations – in fossil fuel-fired power stations. But emissions from household electricity are just the start. Making things warm accounts for almost half of the energy we use inside buildings: globally, 40 per cent of households stave off the chilly weather with central heating systems; raising the temperature of water for showers, boiling pasta, and brewing tea requires energy, too. About two-thirds of this household heat is drawn from fossil fuels, and is alone responsible for 11.5 per cent of global emissions.[8]

Transport is energy-intensive, too. By directing the flow of energy into motion, we move products and people. We haul goods from suppliers to customers, and freight food from fields to supermarkets. A voyage that took our forebears 6 months by ship takes less than 24 hours by plane. It takes me almost 3 hours to walk to work or just over 30 minutes to cycle; if the horizontal English rain defeats me, I can hop in the car and arrive in 15 minutes, snuff-dry. But all this convenience comes at a carbon cost, because the transport sector derives more than 90 per cent of its energy from oil.

Transport alone accounts for 22 per cent of global emissions. Everyday road traffic accounts for almost all of it: 16 percentage points. Whilst individual flights are carbon-intensive, not many people fly, and so aeroplanes contribute just 2.1 percentage points (about the same as ships) to global emissions. The remaining wisps of carbon dioxide come mostly from trains.[9]

And finally, there's the giant cog that turns agrarian societies into developed nations and hauls people out of poverty: industry. It accounts for 24 per cent of global emissions. Manufacturing the material wealth that distinguishes modernity from pre-history is rather carbon-intensive.

The steel industry alone, which fabricates the metallic pillars on which modern infrastructure rests, accounts for 7 per cent of carbon emissions. And the chemical industry — which synthesises everything from the medicines that mend us to the fertilisers that feed us — accounts for another 4 per cent.[10]

So, there we have it: aggregating across all sectors, electricity and heat, buildings, transport, and industry account for 87 per cent of all carbon emissions. The remainder comes from miscellanea such as fishing, agriculture, and public services.[11] All are utterly dependent on fossil fuels. To achieve net zero, we must purge fossil fuels from them all, not just our electricity supply.

Offshoring emissions — such as buying steel from foreign lands rather than making it at home — doesn't count. It goes against the spirit of net zero entirely by shuffling carbon emissions onto the books of other nations.

Reinvention

Net zero would have been impossible until recently. But in our inventiveness, we can take existing technologies and remake them so they run on electricity. This process is called *electrification*. Crucially, we have the option to generate electricity from emissions-free sources, making electrification paramount to net zero.

Electric cars are a good place to start. The idea is simple: by replacing combustion engines with electric motors, we avoid the need to burn petrol and diesel. Governments are promoting them eagerly as part of their net zero strategies. If you don't drive one now, chances are it won't be long until you do, because policies banning the sale of new petrol and diesel cars in the next decade or two are becoming commonplace.

But here's the rub: electric cars don't reduce emissions *per se*. Whilst they're 'zero-emissions' on the road, and aggressively marketed as such, electric cars are only *truly* zero-emissions if the electricity used to charge them is generated emissions-free. Charging electric cars using electricity generated with fossil fuels replaces the *internal* combustion engine with an *external* combustion engine. It defeats the purpose entirely.

When it comes to electrification, it's easy to be distracted by the hype. The point isn't to electrify society for the sake of it, but to electrify society using emissions-free electricity.

Electric cars are just the beginning. We can purge gas from our kitchens by cooking with induction hobs and electric ovens, and by boiling water in plug-in kettles. But decarbonising central heating systems will have the largest gains. The International Energy Agency estimates that the widespread rollout of heat pumps – which run on electricity – could reduce annual carbon dioxide emissions by 500 million tonnes by 2030. That's equivalent to taking every car in Europe off the road.[12] Government support for heat pumps is high in climate-concerned nations, and so if you don't already have a heat pump in your house, it's possible you will within the next decade or two.

As with electric cars, induction hobs and electric heating systems aren't ends in themselves. They're means to net zero. For them to sensibly replace gas and oil boilers, we must power them using emissions-free electricity.

What about electrifying the steel industry? Alas, it's more

complicated than electrifying driving and tea-drinking because three-quarters of the steel industry's energy and feedstock demands are currently met by coal.[13]

Coal offers the steel industry more than mere heat. Iron ore straight from the ground is a combination of iron and oxygen (Fe_2O_3). We chemically pry the oxygen (O) from the principal component of steel – iron (Fe) – by reacting it with the carbon (C) in coal. But those oxygen atoms don't vanish into thin air. Or rather, they *do* vanish into thin air, just inconveniently accompanied by the coal's carbon atoms in the form of carbon dioxide (CO_2).

New technologies promise to decarbonise steel production *electrochemically*. We can, in principle, dissolve iron ore in a solvent, heat the concoction in an electric furnace, and then wrest it into its constituent elements using an electric current. This is, incidentally, similar to how aluminium is made, only at much greater temperatures. Carbon literally wouldn't enter the chemical equation. If it's realised to its full potential, electrochemical steel would effectively replace coal with electricity. It could produce a stream of molten iron that's entirely emissions-free (with the usual caveat that the electricity is generated using emissions-free sources).

Meanwhile, we can already recycle steel using proven technologies in *electric arc furnaces*. One-quarter of the steel in the world is made in this way.[14] As their name suggests, these furnaces run on electricity, and so – if the electricity is emissions-free – they don't emit carbon dioxide. By increasing steel recycling rates, we could offset primary production and thereby reduce steel emissions with minimal effort.

But not everything can be reinvented and electrified. It's difficult to see how we could electrify lorries and other heavy goods vehicles, because of their weighty cargoes and prolonged drive times. Electrifying ships and planes will be even more challenging. And that's before we get started on the chemicals industry.

How do we decarbonise the technologies that electrification can't reach? Those gaps could be filled by the simplest and lightest of all the elements: hydrogen.

Metal and meals

Hydrogen gas is touted as a wonderfuel of the future because, like oil and gas, we can move it through pipes and store it in bottles. And, like oil and gas, we can burn it in boilers and combustion engines.

With hydrogen central heating systems, we could purge gas from buildings where heat pumps either won't fit or won't provide sufficient warmth. We could use it to drive trucks and propel ships, too, because refuelling them could be as quick and easy as refuelling them with oil. And, by concocting *synthetic aviation fuel* using hydrogen as a base ingredient, we could even decarbonise flying.

Crucially, there's a convenient chemical quirk when we burn hydrogen. When we set oil and gas alight, the atoms rearrange themselves into water and carbon dioxide. But when we set hydrogen alight, the atoms rearrange themselves into *just water*.

Instead of using carbon to make steel, we *could* pry the oxygen from iron ore using hydrogen gas instead of fossil fuels. The by-product of combining oxygen with hydrogen is also . . . water. Deployment of this technology is slow. But properly realised, hydrogenated steel could eliminate emissions entirely from the global steel industry, thereby reducing humanity's total emissions by 7 per cent.

We can also use hydrogen to decarbonise the dominant source of carbon dioxide in the chemical industry: ammonia. We make almost all ammonia using the *Haber process*: we take the hydrogen atoms in fossil fuels – like methane gas, CH_4 – and shuffle them onto nitrogen atoms (N) to make ammonia (NH_3). But what happens to the carbon? Unfortunately (and perhaps predictably), it mingles with oxygen to form the dreaded carbon dioxide. Pretty much all the ammonia in the world is synthesised using fossil fuels as its base ingredient, and it's responsible for 450 million tonnes of carbon emissions annually.[15]

But it isn't for nothing. Since the early 1960s, the human population has ballooned by 165 per cent, and yet we grow 30 per cent *more* food per person. To yield such bountiful harvests, we use 800 per cent more ammonia-based fertilisers.[16] The Haber process – one of the most important scientific discoveries of the twentieth century – nourishes the world by effectively turning fossil fuels into food. It saves billions of people every day from hunger. Without it, there would be famine on an unimaginable scale.

To achieve net zero, we must eliminate carbon emissions from ammonia production. But it can't come at the cost of our food supply. Thankfully, we don't face a choice between net zero and starvation. By using hydrogen gas as a feedstock instead of fossil fuels (and by driving the Haber process using emissions-free electricity), we can banish fossil fuels entirely. Using hydrogen, we can decarbonise our harvest and eat it.

The problem with all these lofty aspirations, though, is that hydrogen is difficult to get hold of. Almost none of it is freely available. We hardly ever find it in its pure, elemental form. It's always bound up in some molecule or another, such as water or methane. And therein lies hydrogen's dirty secret: globally, we generate 99 per cent of it using fossil fuels. The associated carbon dioxide emissions exceed 900 million tonnes annually, which is more than is emitted by aeroplanes. For hydrogen to carry weight in achieving net zero, we need to generate it without fossil fuels.

Ironically, given that hydrogen will act as a stand-in for electrification, one of the best ways to generate hydrogen cleanly is using electricity: we can tear water into its constituent elements – oxygen and hydrogen – using a process called *electrolysis*. If the electricity used to perform this chemical laceration is generated emissions-free, then the hydrogen earns the 'green' epithet. Right now, a meagre 0.1 per cent of the world's hydrogen is made this way.[17]

Electricity, electricity, electricity

The electrification of society and the green hydrogenation of industry will make our electricity demand surge. In the International Energy Agency's *Net Zero by 2050* roadmap, global electricity demand increases by more than 150 per cent between now and 2050.[18] It could go even higher, especially when we consider the rapid expansion of electricity-intensive AI technologies. We must generate this new electricity from emissions-free sources whilst simultaneously decarbonising our existing electricity supply. That sounds daunting enough on its own. But it's not *just* decarbonisation we have to be mindful of.

Phasing out gas in favour of electricity risks making people poorer. Kilowatt-hour for kilowatt-hour, electricity in Europe is 1.8 to 3.6 times pricier than gas.[19] It's cheaper to boil water on a hob than in a plug-in kettle, which is why, aside from its charming whistle, my wife and I use a stove-top kettle on our gas-fired cooker. We therefore have to make sure this emissions-free electricity is affordable.

The price of emissions-free electricity will also impact the price of green hydrogen. This will decide the affordability of emissions-free steel infrastructure and the emissions-free ammonia that will fertilise our crops.

And it's vital this emissions-free electricity is reliable. Generating it in massive quantities without using fossil fuels – cheaply and reliably – will determine if net zero is a success or a failure.

So, how do we do it? Two options present themselves: renewables and nuclear power.

The UK is the best place in Europe, and one of the best places in the world, to harness emissions-free energy from the wind.[20] This island nation has an extensive coastline, is surrounded by shallow seas, and is buffeted by prevailing south-westerlies. As such, it makes a natural laboratory in which to test the viability of powering a developed nation using wind turbines.

The political appetite for wind turbines is strong. Raising more of them tops the UK's 'Ten Point Plan for a Green Industrial Revolution'. The prime minister who presided over the plan's conception – Boris Johnson, a self-described wind turbine 'evangelist' – mused in 2020, 'As Saudi Arabia is to oil, the UK is to wind: a place of almost limitless resource.'[21] How's the experiment going?

The UK's wind capacity increased by 460 per cent between 2010 and 2023 to 30 gigawatts, which is enough (on paper) to satisfy the annual electricity needs of 56 million (out of the 68 million) Brits.[22] That sounds impressive.

But there's a catch. *Capacity* is the electricity that would be produced under ideal conditions. For wind turbines, 'ideal conditions' is a euphemism for 'wind that always blows strongly', which, of course, it doesn't.

Generation – how much electricity is *actually* produced – is what really matters, and it falls far below capacity. Britain's wind capacity in 2023 may have been 30 gigawatts, but average generation was only 9.4 gigawatts.[23] The mismatch between generation and capacity is called the *capacity factor*, and we calculate it by dividing the former by the latter. The capacity factor of wind in 2023, therefore, was 9.4 ÷ 30, which equals 31 per cent.

Wind power works best offshore. Places with precious little coastline – like landlocked European nations and US states – have even worse capacity factors. It's a mere 23 per cent in the European Union, which is roughly the same as the world average of about 24 per cent.[24]

Things were particularly bad in 2021. Despite the UK's 5 per cent increase in wind capacity compared to the previous year, there was a 14.5 per cent *drop* in wind generation. The European Union suffered similarly; wind generation fell by 3 per cent despite a 6 per cent rise in capacity.[25] The reason for this power dip is simple: 2021 wasn't very windy. When the wind stops blowing, turbines stop spinning. Capacity is meaningless in the face of calm weather.

There are short-term swings in wind energy, too. Many in the British media celebrated the record-breaking 21.9 gigawatts of wind power generated for a brief period on the blustery morning of 21 December 2023. Many failed to mention that the morning after, wind generation plummeted by about a quarter. A few weeks later, a still period reduced wind power to less than 10 gigawatts for more than 4 days.[26]

It's obvious that wind turbines are at the mercy of meteorology, but it's a point worth emphasising because it exposes the Achilles' heel of wind power: it's unreliable. Wind cannot deliver electricity on demand. It's a shortcoming that has not been satisfactorily addressed by advocates of wind power.

And things become more precarious when we look to the future. Our reliance on electricity will deepen as we electrify society. If there's a blackout *now*, at least most people could still fire up their gas-powered central heating systems and drive their oil-powered cars. What will happen during a blackout when we're all warming our homes with heat pumps and driving electric cars? Society would grind to a freezing standstill until the wind picked up.

The security of our electricity supply will also affect the flow of green hydrogen into industry. This will directly impact productivity. It's one thing to tie the amount of hydrogenated and electrochemical steel we make to the whims of the weather, but what about the amount of fertiliser we synthesise? This jeopardises our ability to grow enough food. We can't sacrifice energy security whilst pursuing net zero.

A perfect storm

When the wind drops, we buttress electricity grids with more reliable sources of energy. In the absence of always-on nuclear – or hydro, for the few nations blessed with the right topography – that means falling back on fossil fuels.

During the cold periods of low wind in winter 2022–2023, the UK repeatedly warmed up coal-fired power stations as a precautionary measure. A cold snap in March 2023 saw them fired up for real as wind generation plummeted. The same thing happened a few months later during a summer heatwave: wind generation tumbled, electricity demand soared as people switched on their fans and air conditioners, and so a coal-fired power station warmed up in case of shortfalls. All these power stations had previously been earmarked for retirement, but the government requested that their lifespans be extended in case they were needed.[27]

In September 2024, the boilers at the Ratcliffe-on-Soar coal-fired power station went cold, marking an end to 142 years of British coal power. Propping up unreliable wind turbines with coal is now a thing of the past, but propping them up with gas is still the norm. The UK's grid is dominated by a combination of gas-fired power stations (about 34 per cent) and wind turbines (28 per cent and increasing); when the latter spin down, the former rev up to meet demand.[28]

There exists a perverse dance-off between gas-fired power stations and wind turbines. Gas's serrated profile is a mirror image of wind's. The two choreograph a ludicrous situation where carbon emissions increase in calm weather.

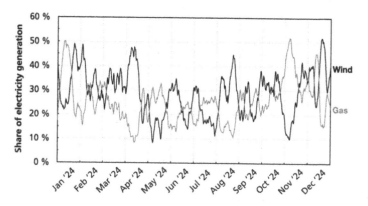

Average wind and gas electricity generation in the UK on a rolling week-by-week basis in 2024.[29]

Wind-heavy electricity supplies – in the absence of a reliable back-up – become unavoidably gridlocked into fossil fuel dependency. There's been a bit of year-on-year fluctuation, but gas's share in the UK's grid has barely budged for a decade.[30] Wind power will not *replace* this gas; it will just *displace* it temporarily whilst ever the wind happens to be blowing.

Things play out similarly on a continental scale. The fleets of turbines the EU built since the dawn of the new millennium have increased wind's share in its grid from almost nothing to about 17.5 per cent. Those wind turbines largely displaced coal power, the dirtiest and least efficient fossil fuel. But they did nothing to help wean the EU off gas. It burns more now than it did in the year 2000, both in relative and absolute terms.[31] It shows no sign of falling. This won't be solved – indeed *can't* be solved – by simply building more wind turbines.

Fossil fuel companies are well aware that over-reliance on intermittent renewables ties us ever more tightly to gas. In 2018, Shell unironically tweeted, 'No sun? No wind? No problem, natural gas has it covered... [gas] is a great partner for [renewables].' BP make it clear that they're playing the same game with tweets such as 'Gas is the perfect partner to intermittent renewables' and 'Our natural gas is a smart partner to renewable energy'.[32] The cynicism of fossil fuel companies increasing our reliance on their product by promoting renewables is breath-taking. The naivety of energy leaders who play into their hands is dispiriting.

Hot air

Whilst promoting solar power in 2017, Elon Musk mused, 'The Sun is the only thing that keeps us from being at roughly the temperature of cosmic background radiation... If it weren't for the Sun, we'd be a frozen, dark ice ball.'[33]

Musk was right on both counts. But unfortunately, neither of these facts has any bearing whatsoever on solar's ability to help us achieve net zero. Such empty rhetoric is frequently deployed by its proponents in a bid to exaggerate claims, obfuscate reality, and compensate for solar's terminal shortcomings.

Solar power works best close to the equator, which doesn't bode well for nations that stretch poleward beyond the tropics. And as the wind drops according to the weather, the Sun drops according to the ceaseless rotation of the planet. Solar panels only work during the hours of daylight, *quelle surprise*, which are in short supply during winter when we need energy the most.

Solar's intrinsic limitations conspire to place the UK second-worst in the World Bank's league table for solar potential after its closest neighbour, Ireland. Of the bottom 30 countries in the World Bank's league table, 26 are European.[34]

Undeterred by the unreliability of solar panels, solar capacity in the EU has grown from essentially nothing to 260 gigawatts over the past 23 years. But remember, *capacity* doesn't keep the lights on; *generation* is what counts, and it's only 28 gigawatts. Despite solar panels becoming more advanced, solar's capacity factor has remained stagnant at about 12 per cent for more than a decade. Those 28 gigawatts supply less than 10 per cent of the EU's total electricity.[35]

Despite the panels it installed in 2021, the UK's solar generation *fell* by 5.9 per cent compared to the year before. Why? The government's Department for Business, Energy and Industrial Strategy partly put it down to 'unfavourable weather conditions', which is a politician's way of saying 'the weather was shit'.[36]

And the dark clouds of geopolitical risk brood on the horizon. China's share in all stages of solar panel manufacturing exceeds 80 per cent. One out of every seven solar panels comes from a single Chinese factory. The International Energy Agency predicts that by 2025, the world supply of solar panels will rely almost completely on China.[37]

Concentrating the conduits of *any* supply chain through the hands of a single nation-state exposes it to exploitation, especially a nation-state that – in the words of former British Prime Minister Rishi Sunak – 'represents the biggest state threat to our economic interests' and 'is a systemic challenge for the world order'.[38]

Solar panels leave our net zero ambitions vulnerable to weather *and* geopolitics. We lean heavily on them at our peril. Didn't Europe learn from the 2022 gas crisis that relying on authoritarian regimes for their energy supply isn't a good idea?

Unstorable

One of the things that makes fossil fuels so reliable is their storability. They persist in geological strata for hundreds of millions of years, and only become useful when we liberate their energy by setting them on fire. Fossil fuel energy can be extracted on demand or stored until we need it.

Wind and solar, by contrast, lack built-in storage. And whilst they frequently fail to generate electricity when we need it, they sometimes generate electricity when we *don't* need it, like during a blustery night whilst most people are sleeping. Couldn't we just store this surplus renewable energy for times of unfavourable weather, thereby smoothing out their intermittence?

Pumped hydro represents 99 per cent of the world's electrical storage capacity. It works by turning dams into batteries: we pump water into an uphill reservoir using surplus electricity, and then release it from up on high through turbines to generate electricity when we need it. If the surplus is generated by wind turbines and solar panels, the dam effectively becomes a giant wind and solar battery.[39]

There are dozens of pumped hydro batteries dotted across Europe. Draining them all completely would power Europe's grid for about 3.5 hours. And then they'd need to be fully recharged (or rehydrated, I suppose). That amount of electricity storage hardly begins to provide enough back-up for wind and solar that can falter for days and weeks at time.

And it's not easy to build more pumped hydro batteries. Most nations don't have the topography to build dams and flood valleys on the scale required. Whilst Norway and Sweden could back up enough electricity for a day or two, thanks to their deep fjords and precipitous peaks, most European countries couldn't manage more than a couple of hours.[40]

What about storing electricity using artificial batteries? Battery technology revolutionised the portable tech industry over the past few decades by enabling luxuries like mobile phones, laptops, cordless power tools, portable speakers, and electric cars. They're also getting cheaper.

Europe's most powerful battery storage facility was connected to the UK's grid in November 2022 in the village of Cottingham. It uses a fleet of Musk's Tesla Megapack lithium-ion batteries – which cost £810,000 each – to store 196 megawatt-hours of electrical energy. That's enough back-up to satisfy the needs of 67,000 Brits for just shy of 5 and a half hours.[41]

But there are 68 *million* people living in the UK. To back up enough electricity for them all for a single day would require 4,500

Cottingham-sized facilities. If, like Cottingham, they used Tesla Megapacks, it would cost about £180 billion – or about 8 per cent of GDP – even before the expenses of installation, maintenance, and end-of-life replacement every decade or two. And a day's worth of electrical back-up is hardly enough given that the weather can change for weeks at a time.[42]

There's a vague hope amongst renewable advocates that batteries will soon be cheap enough to deploy on a scale required to compensate for wind and solar's unreliability. But we need more than hopes. What we really need are *plans*.

What about green hydrogen batteries? In principle, we *could* use surplus wind and solar energy to generate green hydrogen, store it in tanks, and then generate electricity later by burning it in power stations. But in practice, every time we convert energy from one form into another, we incur a loss, which compounds into expensive round-trip inefficiencies. We lose about 70 per cent of the energy when we convert electricity into hydrogen and then back again.[43] It's feasible that hydrogen storage will become more efficient over time, but we don't have a lot of time between now and 2050.

Making a success of net zero requires that we distinguish between what we want to be true and what is true. The dream of a breezy summer's day powering our modern world is just that – a dream. Renewables lack the vigour and reliability to generate the energy we need exactly when we need it. A future of energy scarcity, industrial wind-down, and food insecurity lies ahead should we embrace them.

Wind and solar will never satisfy a society that needs on-demand electricity. Unreliability is a feature, not a bug. Their flow of energy is too erratic, is fiendishly expensive to store, and requires round-the-clock back-up, which locks us into fossil fuel dependency. They are an unserious solution to a serious problem. We need energy today, not promises of jam tomorrow.

That isn't to say that renewable generation is a waste of time. But renewables aren't *the* answer to net zero; they'll only ever be a part of it. Besides, renewables – and the acts of contortion required to compensate for their shortcomings – miss the point completely. The point of net zero is *not* to replace fossil fuels with renewables. The point of net zero is to replace fossil fuels with sources of energy that *don't emit carbon dioxide*.

At best, 'renewable' is a bonus. At worst, it's a distraction that seduces us into reducing emissions in the short term, whilst sacrificing our chances of reaching net zero in the long term.

So, what do we need if we're to reach net zero? Emissions-free sources of electricity. And what do we need if we're to achieve net zero whilst grounding our energy systems in reliability? Emissions-free sources of electricity that are dependable, potent, and on-demand.

Only one source fulfils both criteria: nuclear.

Chapter 4. Net Zero Is Impossible Without Nuclear Power

E minimis maxima
('From the smallest, the greatest')

— United Kingdom Atomic Energy Agency motto

In learning how to split atoms in nuclear power stations, we discovered the solution to net zero *before we learned we were causing climate change*. It's a thought that fills me simultaneously with exhilaration and exacerbation. We have a source of emissions-free energy that could replace fossil fuels... yet we're not implementing it with anywhere near enough vigour.[1]

In 2023, Finland switched on Olkiluoto-3, a modern pressurised water reactor of French design. It's described as 'Finland's greatest single climate act', and for good reason: with a capacity of 1.6 gigawatts, it's the most powerful nuclear reactor in Europe.[2]

So the capacity is large, but what about generation? Well, Olkiluoto-3 – a single reactor – can reasonably expect to generate 12,000 gigawatt-hours of electricity annually, which is enough for around 1.5 million people by our 8 megawatt-hours per person yardstick. That's because nuclear reactors have the highest capacity factor of all power sources. Their global median is 86 per cent. Wind (23 per cent) and solar (11 per cent) are luddite by comparison.[3]

Nuclear power stations churn out electricity on calm, cloudy days, and at night. They don't need propping up with fossil fuels.

We're also good at running nuclear reactors. Almost all their downtime is due to care and maintenance, such as when nuclear engineers swap old fuel for fresh uranium and replace worn parts. These jobs are often planned months or years in advance, allowing us to take power stations offline asynchronously; engineers service one or two reactors whilst the rest of the fleet continues to churn out electricity.

Nuclear electricity is on-demand, too. We can tune nuclear reactors at the turn of a dial – control rods in, control rods out – to generate electricity in tandem with demand. They eliminate the need for impractical storage solutions such as pumped hydro, batteries, and hydrogen.

And, crucially, nuclear power stations do not produce carbon dioxide.

Right now, we could power the *entirety* of the UK's grid with 26 Olkiluoto-3 sized reactors. We'd only need 10 to eliminate fossil fuels from the grid. Even if we kept wind, solar, and hydro generation static – 100 terawatt-hours at last count – and electricity demand doubled, we'd still only need 44 to eliminate fossil fuels *and* make up the new shortfall[4]

On the continent, Europe would need about 170 Olkiluoto-3 sized reactors to eliminate fossil fuels from its grid.[5] Even if its electricity demand *doubled* by 2050, to eliminate fossil fuels *and* make up the new demand, Europe would need to build about 580 reactors.[6] (And that's an unrealistic scenario where Europe stops building wind turbines and solar panels altogether.) Europe built more than 200 nuclear reactors over three decades in the twentieth century.[7] Imagine what it could do now with the added motivation of reaching net zero.

And the whole *world*? To eliminate fossil fuels from the global electricity supply right now, we'd need to build about 1,500 reactors the same size as Finland's Olkiluoto-3.[8] Now let's assume that the world's electricity demand increases by 160 per cent, in line with the International Energy Agency's roadmap. To replace fossil fuels *and* make up this new demand – without increasing renewable generation – the world would need about 5,400 of them.[9] If that sounds like a huge number, that's because it *is*. That's the scale of the net zero challenge. We live in a world utterly dependent on vast quantities of energy. The difficulty of decarbonising that energy is impossible to overstate... but worthwhile endeavours are seldom easy.

In reality, we'd need fewer than 5,400 because the share of wind and solar in grids across the world is increasing. But intermittent sources can't increase indefinitely, and so we'll probably still need thousands of new reactors. If that's what it takes, then that's what it takes. It's time to get building. And yet . . .

We're in the nuclear doldrums. Rarely does nuclear stoke the passions of energy leaders and environmental campaigners in the same way that renewables do. The enthusiasm for uranium is made conspicuous by its almost complete absence. The apathy is even more striking against the backdrop of ever-rising concentrations of carbon dioxide in Earth's atmosphere.

Anti-nuclear sentiments rest on three pillars: expense, environment, and safety. At these three words, you might find yourself nodding slightly. Perhaps you hold one (or all) of these concerns and feel nervous about a massive expansion of nuclear power. These are legitimate considerations. We must rewire our energy systems with caution.

There are valuable lessons to be learned in past mistakes. A good place to start is by looking at what happened in nations that snubbed nuclear whilst trying to decarbonise their energy supply.

Atomkraft? Nein danke.

Few nations are more enthusiastic about cutting carbon emissions with renewables than Germany. It's been embarking on its *energiewende* – or 'energy turnaround' – for more than 20 years. Germany aims to decarbonise its electricity supply by 2035 ahead of full net zero by 2045, and is hailed across the globe as an example to follow. Fatih Birol, Director of the International Energy Agency, said in 2017, '[Germany's] achievements are a big source of inspiration for many countries around the world'.[10] Since 2000, Germany has spent around half a trillion euros on *energiewende*. Total costs will be in the region of €650 billion.[11] What has it got in return so far?

Germany's share of fossil fuels in its electricity supply declined from 64 per cent to 46, and as a share in total energy dropped from

85 per cent to 75. Not everybody is pleased with the pace of progress. The President of Germany's Federal Court of Auditors said in 2019 said that the price of *energiewende* is 'in extreme disproportion to the results'.[12] He has a point: Germany spent half a trillion euros for a 10 percentage point drop in fossil fuel energy.

And it's even worse value for money when you consider that wind turbines and solar panels last for 20 to 25 years, meaning that those built before 2020 will need replacing before Germany's net zero deadline. This, incidentally, is true of all nations that have net zero deadlines close to 2050. By contrast, nuclear power stations tend to last at least twice as long.[13]

In parallel to its *energiewende*, and counter to its net zero ambitions, Germany has also been going through its *atomausstieg*, or 'nuclear phase-out'. In 2010, nuclear power stations generated two-thirds of Germany's emissions-free electricity, which was more than wind, solar, and hydro combined. But just one year later, Chancellor Angela Merkel committed to complete nuclear abandonment by 2022. Her decision was a political one, catalysed by public mistrust of nuclear power following the Fukushima accident in Japan. Between the years either side of Merkel's announcement, nuclear generation plummeted by almost a third. It sank steadily thereafter.[14]

The immediate effect of Germany's rapid nuclear wind-down was an increased dependence on the filthiest of all fossil fuels: coal. In the years following Fukushima, Germany's coal electricity generation jumped by 10 per cent and stayed above 2010 levels until 2017. A 2019 study estimated that Germany could have avoided 300 million tonnes of carbon dioxide emissions between 2011 and 2017 had *atomausstieg* not happened. By 2035, Germany's avoidable emissions may exceed 1.4 *billion* tonnes. For context, Germany's annual carbon dioxide emissions are currently half that.[15]

Coal unleashes more than just carbon dioxide into the atmosphere. It releases massive amounts of airborne particles, as well as noxious gases like NO_x ('nox') and sulphur dioxide. Both of those gases also cause acid rain. The multitude of maladies arising from the air dirtied as a direct result of *atomausstieg* – like heart disease, strokes, and lung cancers – rack up an estimated German healthcare bill somewhere between €3 billion and €8 billion every year. Worse than that, it killed an estimated 5,600 people between 2011 and 2019 alone.[16]

In April 2023, Germany switched off its final nuclear power station. Greenpeace and other anti-nuclear campaigners marked the occasion by throwing a party in the centre of Berlin. Some of them wore canary-yellow *Atomkraft? Nein danke* ('Nuclear power? No thank you') badges. They triumphantly erected a sculpture in the street: a dead *Tyrannosaurus rex* laid atop strewn barrels, each stamped with a ☢ symbol.[17] I'm not sure if they realised that the barrels were *not* the kind we fill with nuclear waste; they were the kind we fill with oil. Their efforts tied Germany closer to the correct contents of the barrels. The irony was lost on them.

In 2010, Germany generated 141 terawatt-hours of reliable, emissions-free electricity in nuclear power stations. That's enough to serve 17.5 million people. Through *atomausstieg*, that generation wound down to nought. During the same period, combined annual wind and solar generation increased by roughly the same amount, meaning all those renewable gains were cancelled out by the nuclear decline.[18] By shutting down reliable nuclear power stations in favour of renewables, Germany also made its electricity supply less secure.

Whilst the UK burned more gas to compensate for a year of calm winds in 2021, Germany burned more coal. In late 2022, the German government extended the lives of coal-fired power stations to boost electricity supplies and regain grid stability. A cold snap in the autumn of 2023 saw part of the Jänschwalde coal-fired power station – which Germany had mothballed 5 years previously – revved back up to prevent blackouts.[19] And throughout all this, Germany – like the UK – made no progress whatsoever in weaning itself off gas. It burns 60 per cent more now than it did in the year 2000. Wind and solar's intermittence must be compensated for somehow.

German politicians knew that rejecting nuclear would make Germany more reliant on fossil fuels, but they did it anyway. In 2014, Sigmar Gabriel – the German Vice-Chancellor – wrote to Stefan Löfven – the Swedish Prime Minister – asking if Sweden's state-owned energy company could press ahead with plans to expand two coal mines in Germany. 'I would be grateful if you could use your influence to make that happen,' Gabriel said. Supplying coal to nearby power stations 'would be important to me personally', he grovelled. His lobbying wasn't even tactful: 'we cannot simultaneously quit nuclear energy and coal-based power generation.'[20]

Germany had a clear choice between coal and nuclear. It chose coal.

In late 2022, reality crashed headlong into anti-nuclear ideology. In a spectacular, and perhaps inevitable, twist, Germany began felling wind turbines to make way for a new coal mine amidst fears of energy shortfalls. A spokesperson for the energy company responsible said, 'We realise this comes across as paradoxical, but that is as matters stand.'[21] In other words: 'deal with it'.

Germany, alongside Sweden, has the tightest net zero deadline of any European nation. But whereas Swedish electricity emits 41 grams of carbon dioxide per kilowatt-hour today, German electricity emits 382 grams. That's because Sweden generates most if its energy from a combination of on-demand hydro (40 per cent), on-demand nuclear (29 per cent), and intermittent wind (21 per cent). And in 2023 – mere months after Germany switched off its final reactor – the Swedish government announced plans to expand their nuclear fleet by as many as ten new reactors. Climate and Environment Minister Romina Pourmokhtari said, 'Expanding nuclear power is one of the most important climate measures for Sweden.'[22]

Instead of spending €500 billion on its failed energy transformation, Germany could have kept its existing nuclear reactors open and spent its money on building more of them. Olkiluoto-3 cost €12.4 billion. Germany could have bought 40 of them for the price of *energiewende*, which would have collectively generated roughly 480 terawatt-hours of electricity annually.[23] With that much electricity, plus the nuclear it switched off since 2000, Germany could have entirely decarbonised its electricity supply, eliminated the need for unreliable wind turbines and solar panels, electrified all 49 million of its cars, and still have spare electricity to generate 1.7 million tonnes of green hydrogen every year, which is 14 times the current global output.[24] And most of those new nuclear reactors would run well into the 2070s.

Simultaneous fossil phase-outs and nuclear phase-outs are incompatible with net zero ambitions. The only way to prop up renewables without getting trapped in a self-defeating spiral of fossil fuel dependence is by building nuclear power stations. The lesson from *energiewende-atomausstieg* is clear: if a nation invests in renewables whilst renouncing nuclear, it will get fossil fuels in return.

Nucléaire? Oui s'il vous plaît!

France and Germany are the two biggest economies in the European Union, they both have legally binding commitments to net zero, and they consume roughly the same amount of energy per person.[25] But they differ wildly in their energy systems.

Of all the electricity Germany generated since the year 2000, 58 per cent flowed from fossil fuel-fired power stations. For France, that number stands at 9.6 per cent. That's because Germany snubbed nuclear and France embraced it. France generated three-quarters of its electricity in nuclear power stations during the time Germany was closing theirs down. As a fraction of total energy consumption, France is the most nuclear nation on the planet.[26] How fitting that it be the home country of our old friend Henri Becquerel, the man who discovered radioactivity.

The 1973 oil crisis saw the price of oil nearly triple within 6 months. At the time, France drew 69 per cent of its total energy from this 'black gold', which placed it at the mercy of the massive price hikes.[27] France decided to do something radical in response. As the saying went at the time: *In France, we don't have oil, but we have ideas.* Theirs were to use nuclear power to free themselves from volatilities in the international fossil fuel market.

Beginning in 1974, France embarked on the biggest rollout of nuclear power that any nation had – and has since – undertaken. It built 54 pressurised water reactors in 25 years. Forty-three of them went live in the 1980s alone.[28] If climate-conscious nations started building nuclear reactors in these numbers and on these timescales *now*, they'd have them in time for 2050.

Whilst it was building its reactors, France simultaneously revved up a home-grown nuclear industry to support them. It built factories to manufacture reactor components. It enriched its own uranium. It fabricated its own fuel rods. And it dealt with its own waste.[29]

In 25 years, fossil fuels as a share of France's total energy plummeted by 36 percentage points, from 91 per cent in 1973 to 55 per cent in 1998. That's an impressive reduction, especially when you consider it happened before technology permitted widespread decarbonisation through electrification and green hydrogen. (Compare that to Germany's 11 percentage point drop over the past 25 years.) At its peak in the mid-noughties, France generated 450 terawatt-hours of nuclear power annually, which satisfied 79 per cent of the population's electrical needs, whereas Germany's wind and solar farms currently generate less than 200 terawatt-hours annually, less than a third of total supply. All 54 of France's nuclear reactors are still working.[30]

France today generates 65 per cent of its electricity from nuclear and only 8.5 per cent from fossil fuels. France has a far lower share of wind and solar in its grid compared to Germany but – kilowatt-hour for kilowatt-hour – emits about 85 per cent *less* carbon dioxide. It's got the second-least carbon-intensive electricity in the EU after Sweden.[31]

But recent years have tarnished France's nuclear record. It hasn't connected a new reactor to its grid since 1999, and the share of fossil fuels in its electricity supply has hovered stubbornly around the 10 per cent mark ever since. Emergency repairs in 2022 – which were already delayed by the COVID-19 pandemic – saw half of its reactors taken offline, which cut overall nuclear generation by almost a quarter.[32] The timing of this nuclear power-down couldn't have been worse, because it coincided with the 2022 gas crisis.

Meanwhile, France increased its combined wind and solar generation ninefold since 2009. That isn't a bad thing *per se*, but those wind turbines and solar panels need propping up somehow, and lacking new nuclear, gas generation increased by more than 50 per cent. Electricity became more carbon-intensive.[33] It's like a watered-down version of the *energiewende-atomausstieg* debacle: investing in renewables whilst neglecting nuclear tied France closer to the fossil fuels it worked so hard to renounce in the 1970s and 1980s.

Despite the stagnation of the past two decades, France's nuclear programme is a stellar success. It's the largest deployment of nuclear technology in the history of any nation, and it demonstrates that it can be rolled out *en masse* if there is a will to make it happen.

And just as it seemed that twilight was falling over France's fleet, President Macron announced in 2022 that the time is ripe for a *renaissance nucléaire*. To reach net zero, France will do what Germany did not: it will simultaneously invest in renewables *and* nuclear. 'We have no other choice but to bet on these two pillars at the same time,' Macron said, addressing spectators beneath the giant metallic blades of an electricity turbine.[34]

I'm not the betting type, but if I were to have a flutter on which major European economy was going to reach net zero by 2050, I'd put my money on France. It expanded emissions-free energy production on the scale required for net zero once before – why not again?

France will begin constructing a new fleet of at least 6 – and perhaps up to 14 – light-water reactors before 2030. There's talk of a French-led 'nuclear alliance' between a dozen European countries to promote a nuclear net zero and oppose anti-nuclear coalitions. 'I have told our German friends and partners,' France's Finance Minister told business leaders in 2023, 'nuclear is an absolute red line for the French government.'[35]

True, France's fleet is ageing; it has an average age of almost 40 years. But there are already more than 150 reactors in the world today that exceed 40 years of age, and France plans to keep its fleet running, churning out emissions-free energy far beyond the middle of this century.[36]

The price of everything but the value of nothing

France spent around €250 billion (inflation-adjusted to 2017 prices) on its nuclear programme and increased its annual nuclear electricity generation by 435 terawatt-hours; Germany spent more than

€500 billion on *energiewende* since 2000 and increased its annual renewable electricity generation by just 190 terawatt-hours. That's twice the money for less than half the energy.[37] And France's nuclear reactors will last at least twice as long as Germany's solar and wind farms.

Between 2015 and 2023, annual electricity generation from renewables increased by 2,800 terawatt-hours globally, at a cost of $4.1 trillion. To increase annual electricity generation by the same amount using Olkiluoto-3-style light-water reactors would have been a trillion dollars cheaper.[38]

If nuclear did make electricity more expensive, then one could reasonably expect countries with a high share of nuclear in their electricity mix to pay more for it. This isn't the case. French households pay less for their electricity than the European Union average. And of the 11 other EU member states that generate nuclear electricity, all but two pay less than the EU average for it.[39]

But aren't renewables getting cheaper and cheaper? It's commonly believed that they are. It's a mantra voiced by journalists and politicians. Even the United Nations in 2022 published a press release titled 'Renewables: Cheapest form of power'.[40]

Arguments such as these are usually based on the *levelised cost of electricity*, an attempt to quantify the price of electricity from different sources. This number comes with a large margin of uncertainty, but it's a simple enough calculation conceptually: first, add up how much it costs to build, maintain, and run a power source; then, add up how much electricity it will generate over its lifetime; and finally, divide the former by the latter. You can thereby compare sources, like-for-like, in cost per megawatt-hour.

A widely read and frequently cited report is Lazard's annual 'Levelized Cost of Energy Analysis' (it makes for riveting bedtime reading). It calculates nuclear is the second-most expensive source of electricity after rooftop solar panels, somewhere in the $221–141 per megawatt-hour ballpark. That's far more expensive than offshore wind ($140–72 per megawatt-hour), onshore wind ($75–24 per megawatt-hour), and solar farms ($96–24 per megawatt-hour).[41]

And that gap widens every year because the levelised cost of renewable electricity *is* falling. The International Renewable Energy Agency details the cost of renewable electricity in an annual report (more gripping reading). The levelised cost of offshore wind has fallen by 63 per cent since 2010. Onshore wind got even cheaper, dropping in price by 70 per cent. Solar got 90 per cent cheaper.[42] Doesn't this dispense with our fantasies of a nuclear-powered future? No, because the levelised cost of electricity misses the point. It quantifies the *price* of electricity but fails to consider its *value*.

The purpose of an electricity source isn't simply to churn out megawatt-hours. It's to churn out megawatt-hours exactly *when* they're needed, *where* they're needed, and in the *quantities* they're needed. A megawatt-hour of electricity that always arrives precisely and promptly is worth more than a megawatt-hour of electricity that may never materialise.

The levelised cost of electricity does not account for the value of reliability or the cost of intermittence. It accounts purely for generation, which is a clanging oversimplification. It also ignores the cost of transmission lines and back-up.

It's tough to put a true price tag on a megawatt-hour of electricity from a particular source. There's an entire field of academic study dedicated to the science (or is it an art?) of calculating these numbers. Even pinning down which variables to include in the calculation is a subject of debate. But we can, at least, make better estimates than the levelised cost of electricity. By considering things such as intermittence, storability, and reliability, we can estimate the levelised *system costs* of different sources.

Renewables become more expensive when we account for their system costs. Nuclear becomes cheaper. By exactly how much varies from estimate to estimate, but the overall picture stays the same: renewables are, all things considered, more expensive than nuclear.[43]

Things normally get cheaper as they become more prevalent. But not renewables. As their share in an energy system increases, so too do their system costs, crystallising into higher household electricity bills, extortionate running costs for electric cars, and unaffordable green hydrogen.[44] The levelised cost of electricity is a creative accountancy of half-truths.

A closer look at Lazard's annual report reveals a footnote that's often glossed over: the levelised cost of nuclear electricity (an overestimate anyhow) is based on a pair of new reactors – Vogtle-3 and Vogtle-4 – at a single power station in the state of Georgia. They're the first nuclear reactors America has built since the mid-1970s. They were 7 years late and ran $16 billion over budget.[45]

The International Energy Agency recently calculated the projected cost of electricity from different sources, based on data from 243 power plants scattered across two dozen countries. Once the upfront and running costs are totted up, they predict that by 2025, new nuclear power stations – megawatt-hour for megawatt-hour – will have the lowest levelised cost of electricity for on-demand sources.[46] And these projections don't even consider system costs.

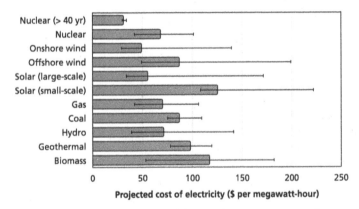

Predicted median levelised cost of electricity by 2025. The upper and lower limits of these predictions are shown by the uncertainty bars.[47]

It *is* true that nuclear power stations are often expensive to build. It's also true they sometimes run over budget; at €12.4 billion, Olkiluoto-3 cost almost €10 billion more than initially estimated. Building a nuclear power station – construction, engineering, financing – typically accounts for 65 to 85 per cent of the cost of nuclear electricity, but once they're built, they're relatively cheap to run.[48] Only nuclear power stations can generate reliable flows of emissions-free electrical power exceeding 1,000 megawatts, which dilutes their large upfront costs.

Longevity dilutes upfront costs even further. The International Energy Agency predicts that a nuclear power station will be the cheapest way to generate electricity by a long way if it lasts for more than 40 years. That is easily achievable. With a mind to extracting maximum energetic value from their existing nuclear power stations, the United States recently extended the lifetime of its fleet from 40 years to 60. In defiant longevity, ten reactors have plans to run until they're 80 years old.[49] I wonder if any of them will become centennials.

Alas, all nuclear reactors reach the end of their lives eventually, at which point they must be dismantled and cleaned up. And during their long lifetimes, they generate nuclear waste. There exists a strong ethos within the nuclear industry that it should pay for its own clean-up, the so-called 'polluter pays' mantra. Many nuclear nations impose a small surcharge on each nuclear megawatt-hour that's generated and deposit it in a nuclear waste fund. It's not a huge surcharge, normally about $5 per megawatt-hour. Keep in mind an average OECD citizen uses 8 megawatt-hours of electricity per year, which amounts to a $40 annual surcharge per person if all their electricity came from nuclear. Nuclear power stations generate a *lot* of megawatt-hours, and so the loose change piles up. This money covers the cost of clean-up. Society, therefore, isn't landed with an eye-watering bill at the end of a power station's lifetime.[50] Nuclear is the only form of power generation that fully internalises the cost of its own waste. Fossil fuels externalise much of their waste in the form of seismic health impacts and environmental damage.

Time is of the essence

Another argument commonly touted by anti-nuclear campaigners is that nuclear takes too long to build. The average build time – from construction to commercial operation – for a nuclear reactor is 9 years. But the *mean* average is overly pessimistic because it's skewed by the small number of reactors that overrun. Therefore, the *median* build time is a more appropriate measure. It stands at 6.4 years. That's not long considering how few nuclear reactors we need to generate an enormous amount of emissions-free electricity. With decades to spare until the net zero deadline, most nations have plenty of time to build a nuclear fleet ahead of schedule.[51]

I share public frustration with the delays in building new nuclear power stations. Hinkley Point C – the UK's first nuclear reactor since 1995 – is at the time of writing still a construction site; it was supposed to take 7 years to build, but it will likely take at least 12. Flamanville-3 – France's first nuclear new build since 1999 – is already a decade late; it was supposed to take 5 years to build but (as I write) is still under construction after 17 years. Delays cost money. A recent analysis in the journal *Energy Policy* showed the longer a nuclear power station is delayed, the more expensive it becomes.[52]

Beset by delays, Hinkley Point C is likely to run at least £5 billion – maybe as much as £8 billion – over budget. The price of Flamanville-3 has more than *quadrupled* since the first estimate in 2004. Delays and price hikes like these make nuclear power look farcical. They shake the confidence of politicians, investors, and citizens, which may scupper our chances of ever reaching net zero.[53]

But these issues are not unique to nuclear. Economist Bent Flyvbjerg describes the 'iron law of megaprojects': *over budget, over time, over and over again.* Delays to all kinds of massive infrastructure projects – France's Concorde programme (1,100 per cent over budget) to Australia's Sydney Opera House (1,400 per cent over budget) to the Suez Canal (1,900 per cent over budget) – are the norm, not the exception.[54] In relative terms, Hinkley Point C doesn't fair too bad at 'only' 30 per cent over budget.

Globally, Hinkley Point C and Flamanville-3 are outliers: from decade to decade, the global median construction time has bounced around between 5 and 7.5 years since the 1950s. We're actually faster now than we were in the 1970s and 1980s.[55]

Through the 1970s and 1980s, Europe's median build time was 7 years and 2 months. France built particularly quickly (6.5-year median), especially considering it built so many. Or perhaps France built quickly *because* it built so many. It's tempting to tinker with technology, but there comes a time to implement the science, put the knowledge to use, and let the engineers build. That's exactly what France did. It picked a standardised reactor design and stuck with it.[56] It essentially built the same pressurised water reactor over and over again. Marcel Boiteux, who oversaw the rapid expansion of France's nuclear fleet as CEO of Électricité de France (EDF), recalled in 2009:

> Every time an engineer had an interesting or even brilliant idea . . . we said: 'Okay, you put this in a file, and it will be for the next series, but now we don't change anything.'[57]

France resisted building the latest, and in doing so, created the greatest.

Since Chernobyl – which marked the end of the second wave – Europe's median build time dragged to more than 13 years. Decades have passed since Europe built nuclear power stations in earnest. During that time, it forgot how to build quickly. The engineers of the 1980s are long since retired. They took their expertise, experience, and well-placed stubbornness with them.[58]

But fast build times persist further east. China built 51 nuclear reactors since 1990, with a median build time of less than 6 years. Across the Yellow Sea, South Korea has built 27 nuclear reactors since the 1970s, which now supply more than one-quarter of its electricity. Its decade-on-decade build time is under 6 years, although a blot on its record saw its three newest reactors take more than 8.[59] Those rapid timescales translate into cheaper power stations. Whilst the cost of nuclear power escalated in Europe over the past 40 years, in South Korea it got cheaper.[60]

More recently, there was a nuclear newcomer on the world stage: the United Arab Emirates. In 2019, it didn't have any nuclear power. By 2024, nuclear generated a quarter of its electricity. All four of its reactors sit in the Barakah Nuclear Power Plant – from the Arabic بَرَكة, meaning 'blessing' – and took about 8.5 years each to build. That's not bad for a first foray into atom splitting. The reactors are of South Korean design; the UAE and South Korean flags fly side by side outside the power station, testaments to the collaboration between the two nations. As the UAE's population grows and its industrial base expands, it's considering building more reactors to help meet its rising electricity demand.[61]

In 2023 I had the pleasure of meeting a bunch of nuclear engineers who work at Barakah at the COP28 climate conference. Many of them were my age. Their enthusiasm was infectious. Barakah has instilled within them an optimism for the future. It is perhaps the most aptly named nuclear power station in the world.

Even if nuclear power *was* slow, it's not like renewables are super-speedy. Seagreen, Scotland's biggest offshore wind farm, went live in 2023. It took 3 years and 4 months to build and will generate something like 400 megawatts of electricity on average, although nobody knows exactly how much because nobody can perfectly predict the weather. That's a build rate of about 10 megawatts of power per month of construction time.[62]

China built six nuclear reactors at the Fuqing Nuclear Power Station between 2008 and 2022. They collectively generate 5,600 megawatts of electricity. From spades first entering the ground to the final reactor revving up, it took China 13 years and 4 months. That's a build rate of about 35 megawatts of electricity per month of construction, more than thrice the rate of Seagreen.[63]

Delays to nuclear power stations are bureaucratic and managerial, not technical or scientific. It is not beyond our wit to make them cheaper by building them faster. We landed on the Moon just 8 years after sending the first human to space; we deployed the COVID-19 vaccine less than 12 months after the start of the pandemic; and we can – and *do* – build nuclear reactors on time in some parts of the world. But we need the motivation to do it again in Europe and America.

Flatpack reactors

Modern nuclear reactors are too big to fit on Fermi's squash court. The gigawatt-sized power stations that coax energy from tens of thousands of uranium rods occupy about a square mile. But not all nuclear reactors are huge. Some are tiny enough to fit on board a military submarine.

By taking inspiration from nuclear-powered submarines, which have been propelled by miniature pressurised water reactors since the 1950s, *small modular reactors* are reinventing the nuclear power station. Small modular reactors are made from individual pieces called 'modules'. Modules can be mass-produced, the pieces transported on the back of a lorry or train, then clicked together with minimal fuss into a fully functioning, miniature power station. I think of small modular reactors as the flatpack furniture of the nuclear world. They're far cheaper, faster, and simpler to build than traditional megaprojects. Small modular reactors will combine the potency of uranium with the efficiency of the factory line.

In 2023, the British government pledged £215 million to help develop small modular reactors, and specifically mentions them in *The Ten Point Plan for a Green Industrial Revolution*. France will invest €1 billion in small modular reactor technology over the coming decade. Even Belgium, which has been embarking on a German-style nuclear wind-down since 2003, is taking a second look. It invested €100 million in 2022. (Incidentally, Belgium tweaked its plan to switch off all its nuclear reactors by 2025 in the aftermath of the 2022 gas crisis; it will keep two running until at least 2035.)[64]

Whilst Ikea might make your flatpack furniture, Rolls-Royce will likely be amongst the companies that make the UK's first small modular reactors. Ikea furniture lasts for a couple of years (if you're lucky), but a small modular reactor from Rolls-Royce will last at least 60.[65]

Admittedly, most small modular reactors aren't as tiny as submarine reactors (nor do they need to be), but they're far smaller than traditional power station reactors. Rolls-Royce is designing its small modular reactor to fit on an area equivalent to 5.5 football pitches. Compare this to Hinkley Point C, which will occupy more than 90 football pitches' worth of land.[66]

The footprint of land occupied by small modular reactors means we can build them where no traditional power station could ever fit. We could wedge them into the places that need electricity the most, like industrial parks, technology hubs, or in AI data centres.

And they hold another major advantage. Wind, solar, hydro, and nuclear all produce emissions-free electricity, but only nuclear produces emissions-free *heat*, too. A factory in Switzerland already uses surplus heat from a nearby nuclear reactor to make cardboard, and in 2024, China began piping steam from a reactor to a nearby chemical plant. Steam generated in another Swiss reactor is used to warm nearby homes. Germany channelled excess heat from a reactor into an industrial saltworks before it shut down in 2003. But these are rare examples. Almost all the excess heat from large nuclear reactors goes to waste because they're built so far away from industrial centres. It's an untapped resource.[67]

Small modular reactors could solve that problem. By assembling small modular reactors close to places that need heat, we can put the steady flow of surplus warmth to good use. We could use it to concoct green hydrogen, purify water, and synthesise ammonia. We could keep people's homes toasty in winter. We could fire steelworks, power food and drinks factories, and regulate greenhouse temperatures. That heat would let us decarbonise hard-to-reach industries as we strive for net zero.

I anticipate a future where small modular reactors are nestled in all major industrial centres across the world. Every large town will have one, too, and every city will have several. In northern latitudes, they'll pipe heat into central heating systems; in the tropics, they'll power air conditioners; in arid climes, they'll desalinate water. They'll power foundries, offices, and chemical plants. You might even have one at the end of your street. Nuclear will become routine.

Small modular reactors may be small, but they're still powerful. Rolls-Royce's will generate 440 megawatts, which is enough for 420,000 people annually. Even by 2050, when electricity demand has doubled, you'd only need about 180 of them to generate all the electricity in the UK.[68] And they will be quick to build. The time between first concrete and first criticality will be 2 years. As they pop up here, there, and everywhere, they'll naturally give rise to steep learning curves. But since they're standardised, delays – and the massive expense they incur – become less likely. And they'll only cost £1.8 billion each. (The word 'only' looks comical when it precedes the word 'billion', but it isn't that much money for a power station that will last at least 60 years, produce torrents of electricity, and emit no carbon dioxide.)[69]

The political will to build expansive fleets of giant nuclear power stations in North America and Europe has largely evaporated. The decade or so (and many billions in capital) it currently takes to build is a deterrent. Small modular reactors are more tangible than traditional power stations. It's not unfeasible a fleet could be commissioned and built within a single election cycle.

Hip young start-ups are dreaming big by going small. More than 70 commercial small modular reactor designs span at least eight nations, including China, South Korea, the USA, Russia, and the Czech Republic.[70] A Danish start-up – Copenhagen Atomics – is designing small modular reactors that have the same footprint as a 40 foot shipping container. The entire thing would fit on the back of a lorry. They're designing them so they can be used to generate electricity *and* supply heat to industry. Copenhagen Atomics also sell 'Energy = Prosperity' t-shirts on their website. It's nuclear chic. Small modular reactors have made nuclear cool.

Miniature nuclear electricity generation is still in its adolescence: there are enormous engineering challenges to overcome, and that's before regulators (and the public) are convinced that it's okay to put a nuclear reactor in a car park. A small modular reactor went critical in China in 2022, and a few more are under construction, but no nation has yet achieved the aspiration of mass production. Rolls-Royce will probably switch on its first small modular reactor in the early 2030s. At the time of writing, it's going through a selection process to decide exactly where to put it.[71]

Fusion won't help us

In December 2022, an artificial star flickered briefly on planet Earth. Scientists and engineers at the National Ignition Facility in California sparked a nuclear reaction that broke all previous records.[72]

They used an arsenal of powerful lasers to mimic the nuclear engine in a star's core. The laser shock crushed hydrogen atoms under the pressure of 100 billion atmospheres; it scorched them to a blistering 100 million °C, far exceeding the temperature in the centre of the Sun. Under these extremities, the hydrogen nuclei zipped around with so much kinetic energy that they were able to approach one another despite their electrical forces of mutual repulsion. They practically touched, getting close enough for nuclear forces of attraction to get a foothold and knit them together. It created new atomic nuclei. Hydrogen became helium. The merger released monumental amounts of energy.[73]

This process – called *nuclear fusion* – is like fission in reverse: instead of generating energy by splitting heavy atoms apart, it generates energy by sticking light atoms together.

Nuclear fusion releases a *lot* of energy. It powers every star in the sky and kindles the rays of sunlight in which the solar system basks. Gram for gram, fusing hydrogen could generate four times as much energy as fissioning uranium, and almost 4 *million* times more energy than burning fossil fuel.[74]

And the by-product? Helium, which has no health or environmental impacts. It's so harmless we use it to fill balloons at children's birthday parties. And, crucially, it doesn't cause climate change. Sustaining fusion reactions and using them to generate electricity has been the holy grail of physics for more than 70 years. If it worked, it would give the world unlimited clean energy.

The National Ignition Facility took the world by storm in 2022 because it achieved *energy breakeven*. The lasers fired 570 watt-hours of energy at the hydrogen fuel, and 880 watt-hours were generated in return. That's a net energy gain of 310 watt-hours. At least, that's if you overlook the roughly 100,000 watt-hours used to charge the lasers.[75] Once you account for that, the experiment consumed 110 times more energy than it generated. That's an efficiency of less than 1 per cent. It wasn't energy breakeven at all. Most media outlets failed to mention this inconvenient fact.

Headlines exaggerated the 'transformative moment'. The *New York Times* reported that fusion 'could provide a future source of bountiful energy'; the BBC said that it 'promises a potential source of near-limitless clean energy'; US Senator Charles Schume hailed the 'astonishing scientific advance [that] puts us on the precipice of a future no longer reliant on fossil fuels but instead powered by new clean fusion energy'.[76] Really?

The breakthrough at the National Ignition Facility was a stellar scientific and technological achievement. But we must place fusion in context when we talk about the pressing issues of energy production and climate change.

Even if we ignore the 100,000 watt-hours used to charge the lasers, an energy gain of 310 watt-hours is tiny. It's enough energy to boil water for a dozen cups of tea, or to keep a single fridge running for about an hour. You'd get half a mile from home in an electric car before you'd be forced to drive back for a recharge.[77] Think how far away we are from generating enough electricity to power a house, or a town, or an entire economy. And that's before we begin commercialising it.

Edward Teller – a nuclear physicist who worked on the Manhattan project and pioneered the early work on nuclear fusion – said in 1958:

> The remaining difficulties appear . . . great enough so that it will take several more years before a machine can be constructed that produces more electric energy than it actually consumes.[78]

Not much changed in the intervening decades. The running joke is that nuclear fusion technology has been 40 years away every year for the past 40 years. I think we'll have permanent human habitation on the surface of Mars before we can power the world with nuclear fusion (although I hope I'm wrong). Barring a miracle, it will not solve our energy woes anywhere close to our 2050 net zero deadline. It's a distraction that robs our attention from credible solutions.

The final question I was asked in 2014 by the panel during my PhD interview was: 'If you could crack one scientific or technological problem, what would it be?' It's a good question and one I still enjoy debating with friends. My answer then was: 'Getting nuclear fusion to work.' It's the answer I'd give today. And I do think we'll do it, one day, but not in time for net zero. But thankfully, we have a proven way of generating plenty of emissions-free energy in the meantime: by splitting atoms apart.

At the 2023 COP28 climate conference, 22 national flags from four continents fluttered side by side. Behind them gleamed a white wall, 15 feet tall, bearing the words:

2050 مضاعفة إنتاج الطاقة النووية ثلاث مرات بحلول عام
2023 الإمارات العربية المتحدة، ديسمبر

TRIPLING NUCLEAR ENERGY BY 2050
United Arab Emirates, December 2023

Each flag represented a signatory of the boldest commitment to nuclear power for decades: the Declaration to Triple Nuclear Energy. It was the first time a commitment to nuclear had been made so explicitly at the world's largest climate conference. The atmosphere amongst nuclear advocates in attendance – me included – was electric. A colleague said to me that 'COP28 will go down as the nuclear COP'.

The declaration's first paragraph formally recognises the 'key role of nuclear energy in achieving global net zero'. It goes on to recognise that 'decreasing nuclear power would make reaching net zero more difficult and costly'. It also specifically outlines a commitment to developing and building small modular reactors.[79]

Nuclear old-timers like the UK, France, Japan, and the USA were amongst the 22 signatories. Germany, the largest consumer of energy in Europe, was a conspicuous (but predictable) absentee.[80] There were a few nuclear newcomers too: Moldova, Mongolia, Ghana, and Morocco. None of them have any nuclear power stations, but they want them. The declaration includes a commitment to support 'responsible nations looking to explore new civil nuclear deployment', which (rightly) places responsibility directly at the feet of old-timers to support their ambitions.

Chapter 4: Net Zero Is Impossible Without Nuclear Power

The declaration w a sign that the conversation about net zero is, at last, becoming more serious. 'You can't get to net zero 2050 without some nuclear', proclaimed US Presidential Envoy for Climate John Kerry to the television cameras. 'These are just scientific realities. No politics involved in this, no ideology involved in this.' French President Emmanuel Macron was more candid with his words: 'Nuclear energy is back.' A year later, at COP29, the USA – the world's second-biggest carbon emitter after China – reaffirmed its commitment.[81]

Beyond the buzz, the naming of the declaration is misleading. It is a call to action but *not* a binding commitment to tripling global nuclear energy by 2050; it merely commits to advancing an 'aspirational goal of tripling [global] nuclear energy' by 2050. I won't joke that it was a 'COP out' (that would be too obvious), but saying 'we *will*' and 'we'll *try*' are two different things.

Besides, a threefold increase in nuclear power generation is grossly inadequate. Nuclear provides about 4 per cent of the world's total energy. Fossil fuels provide more than 80 per cent. Nuclear would have to increase *tenfold* just to cut fossil fuel generation in half.[82]

※ ※ ※

Building nuclear power stations is the greatest act we can take towards curbing carbon emissions. If we started building them now and in large numbers, we'd reach our next zero targets ahead of schedule and enjoy a fleet of reactors until they retired around the year 2100. Commitments such as the Declaration to Triple Nuclear Energy show world leaders are considering such a future. If ever there was a time for society to step up, this is it.

The private sector is showing interest, too. In 2024, Amazon invested half a billion dollars in small modular reactors and plans on using them to power its data centres, and Microsoft and Google announced they'll be using nuclear power to electrify their electricity-intense AI data centres. AI relies on data centres – and data centres need constant, reliable electricity. Data centres account for 3.5 per cent of electricity consumption in the US today, and that figure could rise to more than 9 per cent by 2030. AI technologies will send our soaring electricity demands even higher, but nuclear power could meet those demands without the carbon cost.[83]

And beyond fulfilling our energy needs and enabling net zero, nuclear power offers an antidote to the environmental despair that's suffocating society.

A landmark survey recently asked young people aged between 16 and 25 about their feelings towards climate change. There were 10,000 respondents spanning 10 countries. Almost half of them said their daily lives and ability to function are affected negatively by climate anxiety. More than half thought that 'humanity is doomed'; more than half felt 'powerless' and 'afraid'; three-quarters were frightened of the future. Feelings of betrayal and abandonment by adults were rife. One respondent said, 'I don't want to die, but I don't want to live in a world that doesn't care about children and animals.'[84]

And it's small wonder. The language surrounding global warmi... sorry, global *boiling*, has reached fever pitch. In July 2023, António Guterres made world headlines when he announced:

> Humanity is in the hot seat ... Children swept away by monsoon rains. Families running from the flames. Workers collapsing from scorching heat. The era of global warming is over; the era of global boiling has arrived. The air is unbreathable; the heat is unbearable.[85]

Is suggesting that life on Earth is no longer possible the best way to inspire the world into taking positive action? I think not. Instilling despair is not an effective way to bring about a positive change. All it does is saturate society in doom by conjuring apocalyptic visions of the future.

Informed optimism is far superior. Young people deserve reasons to embrace the world and work to solve the challenges that beset it, full in the expectation that their efforts will bear fruit. We ought to encourage young people to *find* solutions, not *demand* them. Instead of depressing young people, why don't we motivate more of them to work in nuclear? It's one of the best ways to contribute to human progress whilst promoting environmental stewardship. The young people of today will be the engineers, secretaries, managers, plumbers, technicians, scaffolders, computer programmers, electricians, cleaners, accountants, and project planners of the mid-century nuclear power stations. They'll be the nuclear chemists and nuclear physicists doing experiments in mid-century labs. They might even be the ones to get nuclear fusion working. We would do well to encourage them to pour their minds and hearts into something worthwhile and constructive instead of drowning them in eco-doomerism.

Now, I'm not naive enough to think that millions of young people will suddenly take a burning interest in working in a nuclear power station (although one can dream). But at the very least, an honest conversation about nuclear power will rally support for nuclear technology. It will turn climate anxiety into meaningful climate action.

Decarbonising the world using nuclear power should be the twenty-first century's Apollo programme. It would crystallise attainable visions of a better future and unite people under a common goal. Those preoccupied by environmental stewardship would get what they want; those preoccupied by economic growth would get what they want, too. Developed nations would continue enjoying energy abundance, and developing nations would get the energy they need to advance their economies, but all without the carbon cost. It would be victory for the environment *and* for humanity. The world is not a zero-sum game; we don't need to choose between prosperity and sustainability.

By fulfilling our need for emissions-free energy, nuclear power would call forth one of the best facets of the human enterprise: using science and technology to advance progress, promote dignity, and reduce the amount of unnecessary suffering in the world. Now there's a future worth striving for.

Chapter 5. The Green Sheen

We must have zero tolerance for greenwashing.

— António Guterres, United Nations Secretary-General, 2022

Greta Thunberg delivered an impassioned speech at the United Nations Climate Action Summit in New York City in 2019, fulminating against world leaders:

> People are suffering. People are dying. Entire ecosystems are collapsing. We are in the beginning of a mass extinction. And all you can talk about is money and fairy tales of eternal economic growth. How dare you!¹

Thunberg had to cross the Atlantic Ocean from her home in Sweden to attend the summit. But traversing oceans normally incurs an environmental cost. In a bid to lighten her carbon footprint, encourage people to live more sustainably, and bolster her green credentials, she didn't travel by plane. She didn't travel by ship, either. Thunberg travelled by yacht. And not just any old yacht: it was the *Malizia II*, a state-of-the-art 'zero-emissions yacht' that generates electricity using solar panels and an underwater turbine. It took Thunberg — along with her father, two skippers, and a cameraman — 15 days to make the 3,000-mile crossing. Boris Herrmann, captain of *Malizia II*, described the yacht as 'the lowest-carbon option to cross the Atlantic'.²

But Thunberg's transatlantic voyage was only 'low-carbon' if you ignore all the flying it took to pull it off. Herrmann flew back to Europe after the crossing. And then a crew of five skippers flew from Europe to New York to retrieve *Malizia II* and sail her back across the Atlantic. And then *another* skipper flew from Europe to New York to sail Thunberg and her father back to

Europe once the summit was over.[3] In the end, Thunberg's voyage produced more emissions than if her and her father had just taken return transatlantic flights. There wasn't anything zero-emissions about it.

There's a name for deceiving people into believing that something is more environmentally friendly than it actually is: *greenwashing*. And it's rife.

Tech giants greenwash their products when they make pledges of environmentally friendly manufacturing, all whilst encouraging people to ditch perfectly functioning goods for the latest model. Retailers greenwash clothes by making grand public statements about how much they care about sustainability, whilst relentlessly promoting fast fashion and throwaway culture. Supermarkets greenwash grocery shopping by selling 'bags for life' that are 'eco-friendly', when in reality you'd need to reuse them dozens, hundreds, or even thousands of times before they environmentally broke even with as many single-use plastic bags.[4]

And environmentalists greenwash renewables when they promote them as the clean energy sources of the future. That myth that renewables are good for the environment is promulgated by politicians, including António Guterres, who in 2022 proclaimed, 'The only sustainable future is a renewable one.'[5] In this chapter, I'll show how renewables are neither – to borrow two words environmentalists love – green nor sustainable.

Electricity concentration

The conversation about greening our energy supply is dominated by our carbon footprint. But what about the physical footprints we leave on the land?

It's inescapable that all energy sources take up space. There are the concrete jungles of power stations, and fields covered with solar panels and wind turbines. There are also the mines from which we extract fuels (fossil and fissile alike), minerals, and construction materials. Then there are the giant factories in which we manufacture parts. Transmission infrastructure and waste disposal take up space, too.

Every source of electricity has a so-called *electricity concentration*: the amount of electrical energy it generates using a given area of land. Electricity concentration allows us to assess the efficiency of land use. The United Nations and Hannah Ritchie from Our World in Data recently compiled the best estimates of these numbers for different sources. These data account for the land area of the power station, the earth turned during mining, transmission infrastructure, waste disposal, *et cetera*. They also consider the reliability of the source to account for downtime and intermittence.

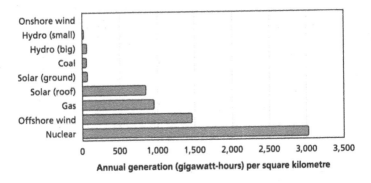

Average electricity concentration of different power sources. Onshore wind is so dilute that it barely registers on this scale.[6]

Nuclear is by far the most concentrated source of electricity. This still holds true when the land area of the Chernobyl exclusion zone in Ukraine (2,600 square kilometres) and the Fukushima exclusion zone in Japan (630 square kilometres) are accounted for (the data in the graph accounts for neither).* Wildlife in both exclusion zones,

---

* They may sound like huge swathes of land, but the Chernobyl exclusion zone is equivalent to a circle just 35 miles across. The Fukushima exclusion zone is equivalent to a circle just 28 miles across.

incidentally, is flourishing. The small amount of land perturbed by nuclear power generation is a major factor in it having the gentlest ecological impact of any energy source.[7]

Nuclear is even more concentrated than wind power from turbines out at sea, far away from the infamous 'my back yard'. Even though offshore wind turbines *themselves* don't take up land area, the mines from which we extract the minerals and materials they're made with do, and they're sprawling compared to the trickles of electricity they generate.

The potency of nuclear fission twinned with the reliability of reactors is why a single nuclear power station like Gravelines in France – which houses six pressurised water reactors on a one-square-kilometre site — can easily power a city the size of London.[8] Even the land occupied by uranium mines and waste management is minuscule relative to the number of gigawatt-hours of electricity it generates.

And small modular reactors are even more compact. Rolls-Royce, for instance, designed its 440-megawatt flatpack reactors to fit on less than 10 acres. With small modular reactors, you could power a city the size of Manchester or Glasgow from a site covering a handful of football pitches.[9]

Renewables, on the other hand, are diffuse. Solar sprawls even in sunnier climes. Greece connected its largest solar farm to the grid in 2022. It occupies 4.5 square kilometres of land and will generate 320 gigawatt-hours of electricity annually. That's enough to meet the needs of 40,000 OECD citizens. The equivalent area dedicated to nuclear power generation would sate the needs of 1.7 *million* OECD citizens.[10]

Rooftop solar panels, as one might expect, have a smaller land footprint than land-based solar panels. Even so, once we account for mining and transmission infrastructure, they use 3.5 times more land than nuclear because of the feeble quantities of electricity they generate.

The picture is more complicated for onshore wind farms. A wind farm generates more power per square kilometre when we squeeze the turbines together, but the slivers of land in between become of little use for much else. Alternatively, we can space wind turbines far enough apart that farmers can grow food beneath their blades, but electricity concentration takes a hit. Either way, onshore wind is dwarfed by nuclear even in its densest form.

More land consumed by renewables means less land for wild spaces and biodiversity. It means less land for forests; for growing food; for thriving flora and fauna; for the natural landscapes that inspire artists and enthral outdoor enthusiasts. If we're not smart about the way we use land, we'll find more of our green spaces replaced by energy infrastructure as we head towards net zero.

There's also an aesthetic component to generating electricity that shouldn't be ignored. It's hard not to feel awed by the sight of a lone wind turbine turning proudly in the breeze, but that's normally because the mines we gouge to make them are in some distant land, far out of sight. Solar farms, in all their glistening modernity, remind me of a dystopian techscape straight from the set of Blade Runner. Nuclear power stations don't look pretty. (Admittedly, I do have a framed picture of Calder Hall in the 1950s on my desk, next to a photograph of Marie Curie.) But they're so power-dense that we'd need tiny numbers to run an entire nation. In their small numbers, they'd be out of sight for the vast majority, and we wouldn't have to desecrate our green and pleasant land.

Mineral rush

COPPER. NICKEL. ZINC. LEAD. CADMIUM. MERCURY.
Those words marked the entrance to a protest camp in Kvalsund, high in the Arctic Circle on Norway's north coast. Nature and Youth – the youth branch of the Norwegian Friends of the Earth – erected the camp in 2021 to protest the construction of a new copper mine. The mine would sit amongst deep fjords, a rolling landscape, and a village of painted wooden houses.[11]

The chemical elements listed on the camp's entrance are just some of the toxic metals that can pollute the environment from mining. Locals and environmentalists fear that the caustic cocktail will contaminate the fjords, disturb the 8,000 reindeer that call the area their home, and defile an area of great natural beauty. Plans for Kvalsund's copper mine were greenlit by the Norwegian government in 2019. The 72 million tonnes of copper ore locked beneath the ground make it the largest copper reserve in Norway. Silje Ask Lundberg, leader of Norwegian Friends of the Earth, decried the project as 'one of the most environmentally damaging industrial projects in Norwegian history'.[12]

The campaigners have a point. Mining has ruinous environmental consequences. It causes water pollution, loss of biodiversity, deforestation, and habitat destruction, and it mires natural landscapes. But mining is also necessary, because much of what makes the modern world turn comes from the rocks beneath our feet. 'If you can't grow it, you have to dig it up', as the saying in economic geology goes. The clash between environmentalists and Norwegian miners is a sign of things to come. The chemical, mechanical, and electrical properties of copper make it a cornerstone in *all* forms of electricity generation. As energy consumption rises – and as we replace fossil fuel-fired power stations with emissions-free alternatives – so too will our demand for minerals.

And copper is just the start. There's an entire slew of minerals that make our energy technologies whirr. The energy sector could easily become the world's largest consumer of cobalt and nickel as we approach 2050. Lithium demand for renewable infrastructure will soar rapidly as we lean more on batteries; I've heard investors call it the 'new oil' and 'white gold'. We'll also have to mine copious amounts of zinc, manganese, and silicon, and neodymium, and chromium (and molybdenum, and graphite, and indium, and praseodymium, and silver, and vanadium, and lead, and . . .). The mineral demand for emissions-free energy sources will at least double by 2050; it could quadruple or even sextuple.[13] Governments must think carefully about how we use these natural resources and strive to dilute their environmental cost by maximising their social benefit.

Recycling will play a role in meeting the surging demand. The more minerals we recycle, the fewer we're forced to exhume. Unfortunately, recycling minerals is (at least for now) technologically troublesome and prohibitively expensive. To compete with mining, recycling must become far easier and cheaper. But even if new technologies enable us to recycle minerals cheaply and at scale, we'll still have to mine fresh supplies to sate our growing demand.[14]

In 2024, the Breakthrough Institute totted up the amount of rock that needs to be excavated to generate a gigawatt-hour of electricity from different sources. They included over two dozen minerals, the material required to make concrete, and – in the case of nuclear and coal – the fuel. The data reveal a stark picture. Coal is by far the most mining-intensive. But wind-power requires 160 to 340 per cent more rock be mined than nuclear; solar 240 per cent.[15]

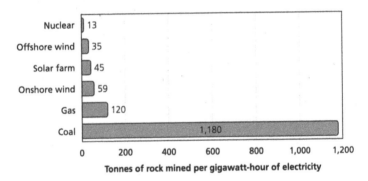

Mining intensity of different electricity sources.[16]

No amount of greenwashing can compensate for the mining intensity of renewables. Sure, they're far less demanding than coal, but they're far more so than nuclear. I find the hypocrisy of environmental campaign groups such as Friends of the Earth, who protest the opening of new mines and simultaneously promote mineral-intense renewables, breathtaking.

There are also humanitarian considerations. Mining minerals is often dangerous, gruelling work. Take the cobalt that's used in lithium-ion batteries, for instance, of which 70 per cent comes from the Democratic Republic of the Congo: in 2022, the US Department of Labor added it to its 'List of Goods Produced by Child Labor or

Forced Labor'. Entire families work side by side in abusive conditions, and when children are left orphaned by fatal landslides, they are compelled to keep digging. Cobalt mining is one of the most tragic exploitations of our time. Demand for this single mineral is projected to more than double by 2050 to meet the demand of electricity storage technologies like grid-scale batteries and electric car batteries.[17]

And then there's geopolitics. Investment in renewables shackles a nation to the security of mineral supplies, far more than investment in nuclear. This is a precarious position, because mineral mines and purification plants are even more geographically concentrated than reserves of oil and gas. When it comes to extraction, Chile mines more than half the world's copper, Indonesia mines more than half the world's nickel, and Australia mines more than half the world's lithium. But one nation in particular holds a disproportionate sway over the mineral trade: China. As well as mining almost three-quarters of the world's critical rare earth minerals, it processes more than half of the world's copper, nickel, cobalt, and lithium.[18] Over-reliance on *any* nation-state for the supply of *any* commodity is unwise; over-reliance on an authoritarian regime for the supply of critical energy infrastructure is foolish.

The carbon cost

I have a confession to make. Every time I've called an electricity source 'emissions-free', I've told a tiny lie. Or perhaps 'a slight oversimplification of the truth' might be a more charitable way of putting it. Nonetheless, describing an electricity source as 'emissions-free' isn't quite accurate because it doesn't consider the whole picture.

It's true that some energy sources don't emit carbon dioxide at the point of electricity production. You won't find carbon dioxide spewing forth from a nuclear reactor. Nor will you find carbon dioxide blowing downwind from a wind turbine or rising from a field of solar panels. But all these electricity sources – every electricity source, in fact – emit carbon dioxide *indirectly* at some points during their lifecycles because energy is required to build, assemble, and run them.

For one, there's the carbon dioxide emitted when we mine minerals and fuels using diesel-powered diggers. Then there's the carbon dioxide we emit when we fabricate a solar panel or make a wind turbine. Power stations, nuclear and fossil alike, are built from emissions-intense steel and concrete using oil-guzzling machinery. Coal is shipped from mine to power station; uranium ore is purified, enriched, and then turned into fuel rods; gas is piped over vast distances to where it's needed. Solar panels and wind turbines are ferried from factory to field. And then there's the carbon dioxide emitted when we manage the waste once they're dismantled. All these processes use energy and so leave a carbon footprint in their wake. Totting up the cradle-to-grave emissions for each power source yields the *lifecycle emissions*.

Nuclear power stations are gigantic pieces of infrastructure, and so, naturally, it takes a lot of energy to build them. Stop Sizewell C – a campaign group trying to halt the construction of a new nuclear power station in the United Kingdom – highlight on their website that constructing Sizewell C will emit well over 6 *million* tonnes of carbon dioxide into the atmosphere.[19]

Stop Sizewell C are correct. Official figures show that construction alone will emit 6.2 million tonnes of carbon dioxide into the atmosphere. But Sizewell C is a 3,200-megawatt power station, and so will generate something like 1,450,000,000,000 kilowatt-hours of electricity during its 60-year lifetime. This means for every kilowatt-hour of electricity it generates, it will emit 4.3 grams of carbon dioxide because of construction.[20]

According to an Intergovernmental Panel on Climate Change data synthesis, nuclear power emits 12 grams of carbon dioxide per kilowatt-hour of electricity once its entire lifecycle is accounted for.[21] And it's by comparing lifecycle emissions with electricity generation that we can place the carbon cost of nuclear power stations in proper context.

Nuclear, it turns out, is about as 'emissions-free' as it gets.

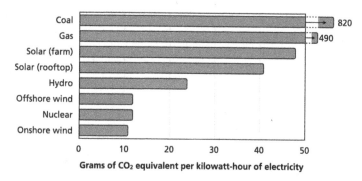

Median CO_2 intensity of electricity sources. Coal and gas are, predictably, off the scale compared to renewables and nuclear.[22]

A big green lie

Biomass – which is just a fancy word for firewood – is often touted as a totally emissions-free source of electricity. It sounds plausible in principle: when we chop, chip, and combust a tree, we incur an emission debt by releasing carbon dioxide into the atmosphere; we repay that emission debt by planting a new tree, which sucks the carbon dioxide back out of the atmosphere as it grows.

The assumption that burning wood is carbon-neutral is held widely; it's enshrined in European Union law – 'The emission factor for biomass shall be zero', which I think has a Biblical ring to it – and exempts burning wood from carbon taxes, whilst allowing it access to lucrative renewable subsidies. A group of 19 nations, which together represent more than half of the world's population, declared at COP23 in 2017 that they would scale up their wood-burning electricity generation as part of their efforts to reduce emissions.[23]

But in reality, wood generates more carbon dioxide per kilowatt-hour of electricity than coal when it's burned. It's a far less efficient fuel. And before we get as far as burning the wood, the machines we use to chop, chip, and dry the trees also emit carbon dioxide. So too do the ships that ferry million of tonnes of the stuff annually from North American forests to European power stations.[24] This adds up to a substantial emissions debt. Wood might *feel* like it's better for the climate than fossil fuels, but a molecule of carbon dioxide has the same impact on the climate whether it came from coal formed hundreds of millions of years ago or from a tree felled yesterday.

Crucially, we only repay the emission debt of burning a forest when – or rather *if* – we allow that forest to grow back in full. That, naturally, is an excruciatingly slow process, taking perhaps a century or more. It's a long time for an emission debt to be outstanding. During the term of the debt, we pay interest in the form of higher concentrations of carbon dioxide in the atmosphere.[25]

There's no guarantee the forests we chop and burn will ever grow back to their natural states; they may never be replanted. Fast-growing plantation forests, which are often cultivated on harvested land, won't solve the problem: they don't soak up as much carbon dioxide as natural forests.[26] They're a vastly inflated currency that will never balance the carbon books.

All these *if*s make it difficult to measure the exact lifecycle emissions of burning wood. Too much depends on whether we allow the forests to return to their natural states and what we do with the land in the meantime: it may be paved over, or the soil washed away by erosion; saplings may be ailed by disease and environmental change, or hewn by loggers before they reach maturity. Either way, the Intergovernmental Panel on Climate Change shows it to be tens to hundreds of times more carbon-intensive than solar, wind, and nuclear.[27]

Wood's inefficiency and long payback times will do little to reduce carbon emissions by 2050. A 2018 study calculated that using wood to offset coal – the most carbon-intensive fossil fuel – would actually *increase* emissions between now and the twenty-second century.[28] This is a critical period, arguably the worst to be adding carbon dioxide unnecessarily to the atmosphere, because we're simultaneously burning fossil fuels at pace.

A 45-minute drive from where I grew up sits the most carbon-intensive power station in the UK: Drax. It doesn't burn coal. Nor does it burn gas. It burns wood. Drax's emissions are so high that even when it's ranked alongside Europe's fleet of coal-fired power stations, it's in the top-ten most prolific sources of carbon emissions. And yet, not a single gram of the 12 million *tonnes* of carbon dioxide it exudes every year counts towards the UK's carbon budget. On paper, they didn't happen.[29]

All 6.5 million tonnes of the trees incinerated at Drax every year come from overseas forest, mostly in North America. Some of those forests are ancient and, until the loggers arrived, offered a sanctuary for rare species.[30]

Drax received £890 million in green subsidies in 2021 alone, and all the electricity it produces is counted as 'renewable'.[31] Counting biomass in this way isn't unique to the UK – all nations lump it under the 'renewable' banner and count the electricity it produces towards their renewable quotas. It's a disheartening example of how energy leaders sometimes take a 'renewables at any cost' mindset at the expense of the very thing they're trying to preserve: the natural environment.

If these trees could talk, they'd be screaming.

※ ※ ※

Renewable is not synonymous with *green*. Renewables can – and *do* – leave an environmental footprint in their wake, even in the absence of enormous quantities of carbon dioxide. Failing to challenge the pernicious myth that renewables are good for the planet is to my mind a major failure of the environmental movement. It's a notion rarely critiqued in conversations about protecting the natural world. Worse than that, it's a delusion promulgated by the same people who claim to act in nature's best interest. When the environmental impact of every car journey and take-out coffee cup is analysed in immense detail, renewables escape scrutiny almost entirely.

All forms of energy generation do some harm to the environment. It's unavoidable. Our collective goal, therefore, ought to be generating the energy we need whilst inflicting the *minimum harm necessary*. Nuclear power meets that ideal.

Alongside the mass rollout of wind and solar, I often hear arguments that we should 'de-grow' our economies and debase our living standards in the name of environmental protection. It's almost always people who live in modern, developed nations who make these arguments, and it's a worldview I don't share. They're arguments in favour of *less*. Less wealth, less energy, less prosperity; less of the progress that characterises the past century. But environmentalism shouldn't make our lives worse. It should make our lives better. Wealth, prosperity, and energy abundance, yes, but *enhanced* by environmental stewardship.

But there's an elephant in the room. A big, awkward elephant. Perhaps this proverbial beast is luminous in your mind's eye. Perhaps it even glows in the dark. Or maybe images of canary-yellow barrels spring to mind, abandoned on some riverbank, lids popped, leaking bright green sludge into the water. The nuclear industry doesn't like to talk about it; anti-nuclear campaigners love to talk about it; I haven't mentioned it much until now, but I've dedicated the next two whole chapters to it. That elephant is *nuclear waste*.

Chapter 6. Nuclear Waste

**DANGER
POISONOUS RADIOACTIVE ☢ WASTE BURIED HERE
DO NOT DIG OR DRILL HERE BEFORE 12,000 A.D.**

A message of foreboding – designed in 1992 by a group comprising an engineer, an architect, an anthropologist, an archaeologist, a linguist, and an astronomer – designed to ward future generations away from buried nuclear waste.

Image courtesy of Sandia National Laboratories.

Nuclear power generates nuclear waste as a by-product. It's an irrefutable fact. 'Nuclear waste' is a nebulous term with many meanings. But at its simplest, it's waste that's radioactive.

And it's not just nuclear power stations that create nuclear waste. *All* nuclear activities do, from the nuclear medicines we administer in hospitals to the uranium that's left behind after enrichment. Some people think this spells an end to our dream of a future powered by uranium. Misgivings about waste are frequently enough to put even the most ardent net zero enthusiasts off the idea of nuclear. The thought gives people the *ick*. In a survey of more than 20,000 people in 2023, 81 per cent of respondents said they were either 'fairy concerned' (34 per cent) or 'very concerned' (47 per cent) about nuclear waste.[1]

Before we write off our best shot at reaching net zero, it's worth thinking about what exactly 'nuclear waste' is.

Half-lives

When radioactive isotopes decay, they sometimes transform into isotopes that aren't radioactive. Radioactive potassium-40 decays, for instance, into non-radioactive argon-40. But some radioactive isotopes decay into *other* radioactive isotopes. Radioactive uranium-235 decays into radioactive thorium-231, which decays into radioactive protactinium-231, which decays into radioactive actinium-227, which . . . and so on, eight more times, until the chain reaches non-radioactive lead-207.* But, eventually, all radioactive isotopes decay to non-radioactive progeny. They just can't stop themselves.

This means radioactive objects naturally become less radioactive with time. And therein lies a strategy for dealing with it: give 'nuclear waste' to Father Time, and he'll turn it into just 'waste'. But how long must we leave it in the hands of our most patient custodian?

Different isotopes decay at different rates. The pace at which a particular isotope transforms into another is unique to that isotope and is set by its *half-life*. Rutherford discovered half-lives in 1900. He noticed some of his samples lost their 'radio-active power' (as he called it) with time; they became gradually less radioactive all by themselves. Eventually, their radioactivity faded so much it became unmeasurable.[2]

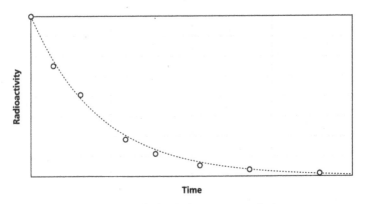

The circles are Rutherford's actual measurements; the decay curve is one I mathematically fit to his data.[3]

---

* The whole series goes: uranium-235 → thorium-231 → protactinium-231 → actinium-227 → thorium-227 → radium-223 → radon-219 → polonium-215 → lead-211 → bismuth-211 → thallium-207 → lead-207.

Rutherford noticed something peculiar about the trajectory of the diminishing radioactivity. It didn't decline in a straight line all the way to zero. It decreased *geometrically*, tracing out a curve that shallowed gradually. He plotted this decay curve on a graph that looked like this:

The decay curve looks exactly like a decay curve I might measure in the lab today, albeit mine would be furnished with more data points and have less scatter thanks to our modern analytical instruments. Rutherford realised that the radioactivity of his samples *halved* with every set interval of time.

The textbook definition of 'half-life' – that you might remember from school science lessons – is *the time it takes for half of the radioactive atoms in a sample to decay*. That definition isn't bad. But it doesn't quite capture the subtlety of what's going on in the atomic realm. How do the atoms 'know' how many other atoms there are in a sample, such that only half of them decay? What if there's an odd number of atoms? What happens when there's only one atom left?

Radioactive decay is governed by the laws of probability. We cannot determine *exactly* when an individual atom will decay. We can only ever know the *odds* that it will decay in a given time. You can think about it in terms of coin-flipping. Imagine a group of atoms, each with its own coin and a fondness for flipping. Heads, the atom wins, and it gets to stick around; tails, it loses, and it's eliminated from the game by radioactively decaying. The coins are perfectly fair, and so the probability of either outcome is equal; it's half-and-half, 50-50.

Now imagine the atoms in the group flipping their coins in rounds, one toss straight after the last. At the end of each round, half – on average – of the atoms decay.

But there's a twist in this game: different types of radioactive isotope wait a different length of time between coin flips. The time between the coin flips is specific to the type of atom. Isotopes that flip their coin frequently will, on average, decay quickly. Those that wait a long time between flips will, on average, decay slowly.

Imagine we have 40,000 radioactive atoms. Let's pretend they're the type that flip their coins once every minute. We begin a perfectly average game. All 40,000 atoms flip their coins. Half of them land on tails; those atoms decay, leaving the lucky 20,000 to progress to the next round. After another minute, the remaining atoms flip *their* coins; half of *them* land on tails, leaving us with 10,000 atoms. After another minute, *those* atoms flip *their* coins; half of *them* land on tails, leaving 5,000 atoms standing. After another minute, 2,500 atoms remain. Another minute, 1,250 atoms. Another, 625.

The *half-life* of these atoms – the time interval between coin flips – is 1 minute.

The 625 atoms would then, on paper, become 312.5. But atoms don't come in halves in practice, and so there's a 50-50 chance of there being either 313 or 312 survivors. And then the game continues.

Inevitably, we'd eventually find ourselves with a single atom. This atom, against the odds, survived. Is it crowned the victor of atomic happenstance? No. It stands alone, flipping its coin once a minute until it too lands on tails and decays. It might only take 1 flip (there's a 50 per cent chance it will); it might take 15 flips (there's a 0.003 per cent chance of that). But it will eventually decay, and join its foregone companions on a different part of the periodic table.

In reality, atoms don't politely flip their coins all at the same time. They flip their coins out of sync with one another. More accurately, therefore, the half-life is the time it takes for half the radioactive atoms in a sample to decay *on average*. Radioactive decay is bitty – it happens atom by atom – but in a mere billionth of a gram of matter there are trillions of atoms. The enormous number of atoms yields smooth decay curves like the one Rutherford plotted.

Half-lives vary from the fleeting – like francium-218, with a millisecond-half-life – to the almost everlasting – like tungsten-180, with a half-life 100 million times longer than the current age of the universe.[4] Now, back to our nuclear waste.

The waste hierarchy

We arrange nuclear waste in a hierarchy of radioactivity. *Low-level* nuclear waste comprises almost all the nuclear waste in the world, including rubble and scrap metal from dismantled power stations. Most of the implements used in nuclear medicine end up as low-level waste, too.

I generate low-level nuclear waste all the time. Almost everything in my lab winds up in the low-level nuclear waste bin: my disposable gloves; most of the vials, pipettes, and test tubes I use in my experiments; the biros I write with in my notebook. Almost all this lab paraphernalia is far less radioactive than a banana, and yet it's still classified as 'nuclear waste' because it came from a nuclear site. In its effort to protect people from radioactivity, the nuclear industry is prudent.

We cram low-level nuclear waste into drums and stack it neatly in trenches or vaults. Its radioactivity diminishes as time passes, and after a century or two, it will have decayed to hardly anything. It's not that long a time; we frequently protect all sorts of objects on those timescales, from books in libraries to paintings in museums.

If you totted up all the radioactivity in all nuclear waste, low-level nuclear waste would account for just 1 per cent of it. It just isn't that much of a problem. That's fortuitous, because it makes up 87 per cent of all nuclear waste by volume.

Intermediate-level nuclear waste is far more radioactive than bananas and biros. We have to contain it in lead and concrete to shield workers from the radiation exposure. Intermediate-level nuclear waste comprises things like the metal dressings that clad the uranium fuel pellets and the graphite blocks used to moderate chain reactions. But it's still a relatively minor concern: it only contains about 4 per cent of all the radioactivity of nuclear waste and comprises just under 13 per cent of its volume.

But when people worry about nuclear waste, they normally mean *high-level* nuclear waste. Despite making up a measly 0.1 per cent of the global nuclear waste volume, high-level nuclear waste contains about 95 per cent of its total radioactivity.[5] And that's a good thing, because it means the hazards are concentrated into a smaller volume of waste, making it far easier to deal with.

High-level nuclear waste consists mostly of spent fuel rods, which hum with decaying atoms. Fresh fuel rods aren't *that* radioactive because the two isotopes from which they're fashioned – uranium-235 and uranium-238 – have half-lives of 700 million years and 4.5 billion years, respectively. Their languorous decay means they emanate radioactivity so slowly it's safe to hold enriched uranium wearing nothing more than cotton gloves. But spent fuel is laced with fission products, which, because of their short half-lives, makes it some of the most radioactive material on the planet.

I've lain my eyes on a bundle of spent nuclear fuel rods before, in Sellafield's Windscale Laboratory. They were the most radioactive and chemically outlandish objects I've ever seen. They resembled charcoal pins, about 1 metre long and as wide as a finger, and were stowed on the other side of a 5-foot-thick window. The glass was ladened with lead, making it transparent to visible light but opaque to beta particles and gamma rays. The glass makes things safe, just like the eclipse goggles that let you look directly at the Sun without frying your retinas. The fuel rods were in a so-called 'cave': a room with 5-foot-thick concrete walls, big enough to fit two vans end-to-end. It was the closest I have ever stood – or will ever stand – to high-level nuclear waste in such quantities, and yet radiation levels were barely above background levels. Sensing my longing for a closer look, a technician picked up the bundle and moved it closer to the window. But he didn't move them using his own hands. Like a puppet master, he manipulated a pair of robot arms sticking out of the cave face, and an identical pair on the inside responded. And when I couldn't help but ask what would happen if I magicked myself onto the other side of the thick leaded window and picked up the spent fuel, he paused for a few seconds before a smile broke across his face: 'You'd be brown bread, lad.'

Fission fragments

Uranium-235 atoms split into two smaller fragments as nuclear fuel rods are showered by neutron sparks. Those fission fragments accumulate over time. Eventually, rods become so choked with fission fragments and sparse in uranium-235 they stifle the nuclear furnace.

They cease to generate sufficient heat to turn electricity turbines. At this point, the reactor's lid is popped, and the spent fuel rods are replaced with fresh ones.* This happens every couple of years, often timed conveniently for the months of low electricity demand.

A medley of fission fragments emerges from splitting uranium atoms with half-lives spanning from split-seconds to many millions of years. The ash from a nuclear furnace contains most of the periodic table, from the 'impossible' barium, discovered first by Hahn and Strassmann, to various isotopes of krypton, caesium, strontium, technetium, and iodine. Most fission fragments have relatively short half-lives, which is why spent fuel rods are far more radioactive than fresh ones. But nuclear fission ceases to occur when we take fuel out of a reactor, and so those fission fragments cascade down a series of decays until they reach the stability of a non-radioactive isotope. The radioactivity of spent fuel, therefore, diminishes with time.

Atoms with short half-lives in spent fuel aren't a problem for very long because they live fast and die young. About 99 per cent of the fission fragments in a spent fuel rod wither within 4 years of being taken out of a reactor.[6] During this period, the fuel rods are stowed in pools of water that in nuclear parlance are (delightfully) called 'ponds'. These ponds, each deep enough to submerge a four-storey building and containing as much water as a dozen Olympic pools, serve the dual purpose of cooling the rods and shielding workers from the radiation.

Longer-lived fission fragments are more problematic. Two of the most notorious are strontium-90 and caesium-137. Spent fuel is chock-full of them. After the short-lived radioisotopes vanish, strontium-90 and caesium-137 account for most of the remaining radioactivity. They each have half-lives of about 30 years, which is in a nightmarish Goldilocks zone: short enough to kick out a tremendous amount of radioactivity over decades, but long enough to remain radioactive for centuries. After about 500 years, though, more than 99.999 per cent of the strontium-90 and caesium-137 will have

---

* CANDU reactors, incidentally, are designed to continue churning out electricity whilst their spent fuel is being replaced. There's no need to switch them off.

decayed. It's a long time, but not incomprehensibly so. There are pubs older than that.

Collectively, the radioactivity of fission products wanes by 99.9 per cent within 40 years; it falls by 99.9999 per cent within 400 years.[7] And so they're only *really* problematic on timescales of about a millennium. But even after the fission products wane, high-level nuclear waste will still be radioactive because of a different class of atom altogether: those that lie beyond uranium.

Beyond uranium

We don't find atoms with atomic numbers greater than 92 in nature. Their non-existence is explained neatly by all isotopes heavier than uranium-238 being radioactive, with half-lives much shorter than the age of our planet. Any such elements that *were* present when Earth formed 4.5 billion years ago have long since vanished. Today, only their progeny remains.

And yet, a closer look at the bottom row of a periodic table reveals the aptly named *transuranic* – 'beyond uranium' – elements. Transuranics don't come from Earth's chemistry set. They come from us. We, humans, made them in nuclear reactions.

In the early 1940s, Fermi and his colleagues were perfecting the science of splitting uranium atoms using neutrons. Meanwhile, different teams of scientists were pursuing an altogether different idea: what if instead of *splitting* uranium atoms, you could make them bigger? What if you coaxed uranium into absorbing subatomic particles? You could, at least in principle, make atoms with more than 92 protons in their nuclei and add a tile to the periodic table.

It was by this method that nuclear physicist Edwin McMillan of the University of California, Berkeley began charting new chemical landscapes. For this work, he would win one-half of the Nobel Prize in Chemistry 1951. In early 1939 – a few months after the discovery of nuclear fission – McMillan placed a sprinkle of uranium atop a piece of filter paper and buried it under a small stack of cigarette paper. He placed this odd contraption in the firing line of a neutron gun.

When uranium-235 bursts apart, it sends fission fragments flying at tremendous speeds. By sequentially peeling away his layers of cigarette papers and measuring the radioactivity of each, McMillan wanted to see how far different fission fragments had penetrated the stack. The faster a fission fragment streamed, the further through the stack it would travel. He thought this would allow him to measure how energetic fission fragments were.[8]

'Nothing very interesting about the fission fragments came out of this,' McMillan recalled 12 years later during his inaugural Nobel Lecture.[9] He found the concoction of fission fragments was the same on each layer of cigarette paper. Yawn.

The filter paper on which the uranium target sat, however, 'showed something very interesting'. On it, McMillan detected radioactivity from two isotopes that *weren't* present on the layers of cigarette paper. This meant they weren't fission fragments. One of the isotopes had a half-life of about 25 minutes; the other, about 2 days. Both emitted beta radiation.

He published his results in a letter to the editor of *Physical Review*. He ascribed the 25-minute half-life to an isotope Meitner, Hahn, and Strassman had discovered 2 years previously: uranium-239.[10] Whilst the light isotope of uranium – uranium-235 – splits when it's whacked by a neutron, the heavy isotope – uranium-238 – sometimes swells. It *absorbs* neutrons and thus grows into a heavier isotope, uranium-239. But McMillan's 2-day half-life was new. What was it? In his letter, he stopped shy of speculating. Privately, though, he strongly suspected it was the radioactive progeny of uranium-239.

Recall: when an atom beta decays, its mass number stays the same, but its atomic number increases by 1 because it gains a proton. It turns into the next element along the periodic table. The decay product of uranium-239, therefore, ought to be an isotope of a brand-new element: the then-innominate element that occupies the 93rd tile.

But it was a tricky hypothesis to test. Blended with a frenzy of fission fragments, it was difficult to tease the radioactivity of the mystery isotope. McMillan, this time with friend and fellow nuclear physicist Philip Abelson, repeated the experiment. Racing against the 25-minute half-life of uranium-239, the pair frantically separated it from the filter paper and chemically purified it from the fission fragments. Then they started measuring. As the radioactivity of their uranium-239 diminished, the radioactivity of the decay product – the isotope with the 2-day half-life – grew. This proved the mystery isotope was the progeny of uranium-239. McMillan and Abelson sent another letter to the editor of *Physical Review*, titled simply 'Radioactive Element 93'.[11]

Uranium takes its name from the planet Uranus. McMillan and Abelson named their new element after the next planet out: *neptunium*, with the chemical symbol 'Np'.

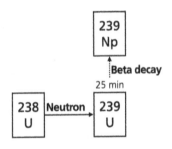

So, why does any of this matter when we consider nuclear waste? It matters because only a couple of per cent of the uranium atoms in nuclear fuel are the *fissile* isotope, uranium-235. The rest are the *fertile* isotope, uranium-238, from which transuranic elements grow. Uranium-238 atoms in nuclear fuel act as seeds that sometimes absorb neutrons and grow into uranium-239, which beta decays swiftly to atoms of neptunium-239. And then those neptunium-239 atoms, with their half-lives of 2 days, beta decay to atoms that occupy the 94th tile of the periodic table.

Element number 94 is named after the former planet just beyond Neptune, which is itself named after the Greek god who presides over the afterlife: *plutonium* (chemical symbol 'Pu'). Nuclear fuel accumulates plutonium when it's showered by neutron sparks. That nuclear reactors synthesise brand new elements in their cores makes them like stars.

Plutonium-239 is an entirely different beast to neptunium-239. It alpha decays with a half-life of *24,000 years*, and there's a lot of it in spent nuclear fuel. Occasionally, a plutonium-239 atom acts as a seed and absorbs another neutron. It grows to become plutonium-240. But this does little to alleviate the radioactivity, because plutonium-240 alpha decays with a half-life of almost 7,000 years.

And to add another element to the mix, plutonium-240 sometimes absorbs a neutron and grows to become plutonium-241. This isotope has the shortest half-life of the common forms of plutonium – a mere 14 years – and radioactively decays by spitting out a beta particle. That means it transforms quickly into the *next* element along the periodic table from plutonium: element number 95, *americium* (Am). (They'd run out of planets by this point.)

Americium-241 is a special isotope. It has a set of properties that make it suited to powering spacecraft. It will enable humanity to explore the outer solar system, rove the surface of Mars, and build habitats on the surface of the Moon. For now, though, we'll let the americium-241 in our spent fuel decay – as most of it does, with a mediocre 430-year half-life – by spitting out an alpha particle. Remember: when an atom alpha decays, its atomic number decreases by 2 and its mass number decreases by 4. Americium-241, therefore, decays to neptunium-237.

Neptunium-237 has an even longer half-life than plutonium-239 and plutonium-240. It's more than 2 *million* years. In all, we end up with a variegated medley of long-lived transuranics, grown by showering fertile uranium-238 with neutrons. A fuller picture of the various transformations reveals a branching spirograph of alchemy.

The nuclear scientists of the mid-twentieth century synthesised plutonium by the millionth of a gram on chemistry benchtops. Nowadays, we make it by the *tonne* inside nuclear reactors. Once it's been inside a reactor for a couple of years, about 1 per cent of a fuel rod is made from plutonium alone. The glacial half-lives of plutonium's major isotopes mean spent fuel will remain radioactive into the far future.

Radioactivity can be counter-intuitive: fission products are, gram for gram, incredibly radioactive, but vanish quickly; transuranics, on

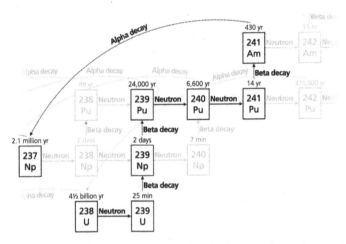

The nuclear reactions that synthesise long-lived isotopes of neptunium, plutonium, and americium. The times above the tiles are half-lives. I've also shown minor nuclear reactions, faded for clarity. There are dozens more nuclear reactions happening simultaneously – plus a criss-cross of radioactive decays – but I had to stop somewhere . . .

the other hand, are less radioactive, but remain so for far longer. It will take about 100,000 years for the radioactivity of spent fuel to decay back to that of the uranium ore whence it was fabricated.[12] That's about the same amount of time that *Homo sapiens* have walked the earth.

Sizing up the problem

It's easy to feel overwhelmed by the enormity of 100,000 years. We're a species that evolved to consider time in terms of decades, not millennia. It's worth taking a step back, therefore, and looking at how much high-level nuclear waste we're talking about.

In addition to 7.5 *billion* kilowatt-hours of electricity – meeting the electricity needs of a million or so OECD citizens – a typical 1-gigawatt nuclear power station generates about 30 tonnes of spent fuel annually. That amounts to about 30 grams per person, the same mass as a few grapes. Over a lifetime – 80 years, say – it amounts to about 2.5 kilograms. And given it's three times denser than solid rock, it doesn't occupy much space. It would barely fill a wine glass.[13] You get a huge amount of energy from a tiny amount of uranium; the upshot is that you get a tiny amount of nuclear waste for a huge amount of energy.

How does high-level nuclear waste stack up globally? The world spawned 390,000 tonnes of it over the past 70 years. All that waste would fit into a cube just shy of 33 metres across. That's it. All the high-level waste – from the entire history of global nuclear power generation – would fit inside a modest-sized concert hall.[14]

We're still accumulating it, of course. The world generates about 12,000 tonnes every year. That makes the 33-metre-wide cube grow by about a foot annually.[15]

The boring truth about nuclear waste is that it just isn't much of a problem. There's so little of it. But even so, the question of what to do with that 33-metre-wide cube that will be radioactive for the next 100,000 years remains. Right now, it's dotted across the world's nuclear sites in ponds and temporary storage. But it can't stay there forever.

Going underground

The only way to *truly* rid ourselves of high-level nuclear waste would be to launch it into outer space. But 1 in every 20 rockets that takes off never makes it, which is an unacceptable failure rate if you're carrying highly radioactive cargo.[16]

Lacking a method to sensibly dispose of nuclear waste astronomically, the plan for essentially every nuclear nation on Earth is to dispose of it *geologically*. That's a euphemism for 'burying it'. It's not disposal *per se*, which implies the waste vanishes. It's more 'long-term storage' whilst its radioactivity wanes.

But how do we mark the whereabouts of nuclear waste and warn future generations they're not to tamper with it? When pondering such things, we must prepare for the possibility – or perhaps the *inevitability* – that civilisation as we know it won't exist in the far future. Our social institutions and cultural inclinations might change. I think it's almost certain they *will* change, drastically, and the whereabouts of the waste will be forgotten. Therefore, it's not as simple as stockpiling the waste in a cave and erecting a 'DO NOT ENTER' sign at its mouth. English won't be read widely in the 1,021st century. Would you turn around if you saw '𓃭𓃮𓄿' carved into a rock? (These are ancient Egyptian hieroglyphs for 'Do not enter!'.)

In the early 1990s, the United States Government invited experts to devise symbols and artworks that would warn future generations. I used one as this chapter's epigraph. But I much prefer Carl Sagan's idea of a universal symbol of foreboding:

> It is the symbol used on the lintels of cannibal dwellings, the flags of pirates, the insignia of SS divisions and motorcycle gangs, the labels of bottles of poisons – the skull and crossbones. Human skeletal anatomy, we can be reasonably sure, will not unrecognizably change in the next few tens of thousands of years.[17]

The skull and crossbones is the archetypical *memento mori*, recognised by virtually every culture that's ever existed. Alas, I suspect even chiselling it into the cave entrance wouldn't be enough to deter people. If anything, it would probably entice them. We just can't help ourselves. Besides, the chances any structure or monument marking the whereabouts of nuclear waste would survive long enough to be of use are next to none.

We have little choice but to store nuclear waste deep underground, where it's unlikely to ever be happened upon. No one has gone further towards achieving geological storage on 100,000-year timescales than far-sighted Finland.

Subterranean Scandinavian

Finland has generated more than 900,000 gigawatt-hours of nuclear electricity since the late 1970s and currently generates

42 per cent of its electricity inside nuclear power stations. It's consequently accumulated about 2,600 tonnes of high-level waste, which would fit inside a cube 6 metres across. The Finnish Nuclear Energy Act legally compels Finland to handle, store, and permanently dispose of it. There's no kicking the can over the border. In that spirit, the Finnish government took decisive action in 1983 by approving plans to entomb spent fuel underground. After combing the landscape for a suitable site, it settled on Olkiluoto Island.[18]

Olkiluoto Island sits just off Finland's west coast, 140 miles northwest of Helsinki. Backed by pine trees on the mainland side and the Baltic Sea on the other, the island spans just 4 miles end to end. Yet almost a third of Finland's electricity comes from this single speck of land because it's home to three of Finland's five nuclear reactors, including Olkiluoto-3.[19]

The reactors are built atop some of the toughest bedrock on the planet. Crumbly sandstones that formed 1.8 billion years ago were baked and pressed over aeons into a hard, watertight geological subsurface. Olkiluoto has, for all intents and purposes, been around forever.[20] It won't be changing drastically any time soon, either, because it's 900 miles from the nearest tectonic plate boundary. It's geologically static.

All this makes Olkiluoto an ideal place to entomb nuclear waste. Excavation of a final nuclear resting place – named Onkalo after the Finnish for 'hiding place' – began in 2004. It's the world's first and only permanent disposal facility for high-level nuclear waste.

Onkalo's tunnels spiral 1,500 feet beneath the Scandinavian landscape. The subterranean passageways, with their twists and switchbacks and branches and offshoots, already burrow through 6 miles of crystalline bedrock. A cave mouth, wide enough to admit a lorry, grants passage to the nuclear catacombs.[21]

Before Finland lays its spent fuel rods to rest, it will submerge them in ponds for a few decades where their radioactivity will decrease a thousandfold. But the long-lived fission products – and, of course, the transuranics – will persist. Once they've cooled off, the fuel rods will be dried by mechanical workers (i.e. robots) in the *encapsulation plant*. This facility has 4-foot-thick concrete walls that shield biological workers (i.e. humans) from the radiation.[22]

The mechanical workers will neatly bundle the dry fuel rods into giant cast iron tubes. Each cylinder, 3 feet in diameter, will be as tall as two or three people standing on each other's shoulders. They'll weigh close to 30 tonnes. Once the fuel is inside, the cylinders will be nested inside a copper jacket. The chemically unreactive copper, 2 inches thick, will protect the steel tube from corrosion should water infiltrate the underground caverns.[23]

The intense radiation emanating from spent fuel is enough to break the molecules of the air – namely oxygen (O_2), nitrogen (N_2), and moisture (H_2O) – into pieces. Those pieces may reform into chemical undesirables, like pressurised hydrogen gas (H_2) and corrosive nitric acid (HNO_3). The air in each cylinder, therefore, will be purged and replaced by inert argon gas.[24] Then, the cylindrical caskets will be welded shut. It will be the last time anybody lays eyes on the spent fuel, ever.

From the encapsulation plant, the canisters will be lowered one by one down a lift shaft. Upon arrival in the catacombs, they'll be greeted by a remote-controlled wagon that will stow them in vertical pits.[25]

Each pit will be lined and plugged with a final geological barrier: clay. When it gets wet, clay expands and warps, and so will tighten its grip on the canister and close any conduits into which water might seep. The self-sealing clay will also stop nuclear waste from getting *out* should any of the canisters leak. Given the canisters are designed to last for at least 100,000 years, though, this is unlikely.[26]

When no vacant pits remain, the caverns will be backfilled with more clay and plugged with a giant concrete wedge. The first fuel will be laid to rest in 2025. It took Finland 40 years to go from approval to implementation.[27]

Sometime in the third decade of the twenty-second century, the 3,000th canister will be sent underground. Onkalo will be full. By then, it will house almost 6,000 tonnes of high-level nuclear waste, including all the spent fuel from Finland's five nuclear reactors over their entire operational lifetimes. (There's a chance its storage capacity might double to 12,000 tonnes, making space for more spent fuel from nuclear reactors yet to be built.) Then, the nuclear necropolis will be sealed for good using more clay. And finally, the surface-level buildings will be demolished. As the millennia crawl by, Onkalo's whereabouts will slip from humanity's collective memory. It's incredibly unlikely our descendants will ever uncover it, because it is but a tiny speck on the surface of the Earth.[28]

Ice ages will creep back and forth over Finland in the coming million years. Miles of ice will weigh heavily on the geological subsurface, but Onkalo has been built to withstand the crushing pressure and rippling seismicity. Geologists positioned the burial site between two parallel fault lines, along which any shifts in the subsurface will preferentially happen. Onkalo is earthquake-proof.[29]

'But what if the canisters *do* leak?', you might reasonably ask. A fatal design flaw may have escaped the notice of thousands of engineers; a freak event, such as an unusually massive earthquake, might rupture the layers of containment. Against all odds, radioactive atoms might leak out of the canister and be swept away by subterranean water.

Even in such an event, the radioactive atoms have a long way to travel before they reach the surface. Some of them, like the caesium-137 – which will be all but decayed within a few centuries anyhow – would be soaked up naturally by the rocks they flow through. All of them would be vastly diluted by the mobilising water. Even under a worst-case scenario, people living near Onkalo who drank from contaminated wells would be exposed to negligible amounts of radiation.[30]

The canisters; the self-sealing clay; the geological good fortune. All these things will conspire to make Onkalo last 100,00 years, 20 times longer than the entirety of recorded human history. The Great Pyramid of Giza isn't even 5,000 years old. Onkalo – on a human timescale – will be around forever.

Onkalo will cost €3.3 billion, including operation until the year 2120. The bill will be covered by Finland's State Nuclear Waste Management Fund, accumulated from the small surcharge added onto the price per kilowatt-hour of nuclear electricity. The longevity and reliability of nuclear reactors mean that globally, managing and disposing of spent fuel accounts for less than 1 per cent of the cost of nuclear power.[31]

Opposition to nuclear is minimal in Finland. In fact, support for nuclear power is at its highest levels since records began in the early 1980s.[32]

Sellafield is one of the most complicated nuclear sites in the world. Large stretches of it were assembled during the 1940s and 1950s in an atmosphere of Cold War haste. Many of its buildings – including some of the world's first nuclear reactors – were raised quickly, giant experiments that were the foundation of many of the nuclear technologies we use today. But the fervour of the atomic arms race, and the political pressure to establish a nuclear programme, left little time for future considerations, and plans for dismantling those facilities weren't a priority. This left a legacy of troublesome nuclear waste. Those pioneers – inevitably – made mistakes along the way. The nuclear engineers of the mid-twentieth century didn't have a rulebook. They were *writing* the rulebook.

Sellafield today is full of nuclear science oddities and unusual, though no more radioactive, types of waste. Cleaning it up represents a thorny engineering challenge. It will take more than a century to fully dismantle. The old nuclear waste at Sellafield and other sites like it is a problem the world inherited from a generation past. *New* nuclear waste, on the other hand, is not a problem. Finland found and implemented a solution by building Onkalo. Geological deposition is the ultimate resolution to the nuclear waste conundrum.

Several nations are emulating the Finnish approach. Sweden will have its own nuclear necropolis by the mid-2030s, and France by mid-century. At least nine other nations have them in various stages of planning. The UK will probably entomb its high-level waste close to Sellafield.[33]

Burying high-level nuclear waste is the 'out of sight, out of mind' method. Part of its appeal lies in its conceptual, practical, and political simplicity. But there is another way. In its straightforwardness, geological disposal fails to capture the subtleties of what spent nuclear fuel is actually made from. It turns out that nuclear waste, far from being a liability we should throw away, is a resource that can be put to good use.

Chapter 7. Nuclear for the Third Millennium

'This is, uhh, this is heavy duty, Doc, this is great. Uhh, does it run,
like, on regular unleaded gasoline?'
'Unfortunately, no. It requires something with a little more
kick . . . plutonium!'
'Ah, plutonium . . . wait a minute. Are you . . . are you telling me
that this sucker is nuclear?!'

– Marty McFly and Doc Brown, *Back to the Future*, 1985

The inbuilt dictionary on my iPhone defines 'waste' as:

(noun) unwanted – or unusable – material, substances, or by-products.

It even uses 'nuclear waste' as the example to epitomise the definition. But how fair is this view of radioactive nuclear by-products? 'Unwanted'? Perhaps. But 'unusable'? Far from it.

Nuclear fuel begins life as unspoiled uranium. That uranium is usually enriched; the fissile atoms – uranium-235 – are elevated from their natural abundance of 0.7 per cent to somewhere between 3 and 5 per cent. The rest of the atoms are uranium-238, which don't participate in fission chain reactions.

After a couple of years inside a nuclear reactor, most of the uranium-235 atoms are broken. They burst into fission fragments and neutron sparks when they liberate their energy. But not all of them. Some never had the chance to fission. Even after a good roasting, there are still plenty of fissile atoms left over at the end of a fuel rod's life.[1]

In fact, spent fuel typically comprises about 95 per cent slightly enriched uranium, 4 per cent fission fragments, and 1 per cent plutonium. (There are trace amounts of neptunium, americium, and other transuranics.) The exact blend varies from rod to rod; it depends on the initial enrichment of the uranium, where exactly it sat in the reactor, and how long it was burned. It also depends on the reactor type. Either way, *spent fuel is still made mostly from enriched uranium.*

Considering this isotopic good fortune, indiscriminately disposing of spent fuel rods wholesale makes little sense. They still contain an enormous amount of usable, emissions-free energy. 'Waste' here isn't a *noun*, but a *verb*, defined by the dictionary on my iPhone as:

(verb) use – or expend – carelessly, extravagantly, or to no purpose.

(Ironically, my iPhone uses the example 'We can't afford to waste electricity' to typify this definition.)

This isn't semantic pedantry on my part. Several nuclear nations regard Finland-style nuclear waste disposal as squandering a valuable resource. Rather than dispense their spent fuel in bulk, these nations make efforts to sort and recycle it. Unsurprisingly, France is a global leader.

Sorting the periodic table

When you recycle a metal can, it can be used again. It might become part of a new jumbo jet wing. It might become part of a new frying pan. The luckiest ones become part of a new nuclear power station.

Recycling household waste only is possible once you've separated it into its various components. The sorting process is the first step towards turning waste into something new. Spent nuclear fuel is more complicated than household rubbish, though, because it teems with radiation and contains most of the periodic table. Therefore, we can't sort it by hand into different coloured bins. We have to sort it using robots in factory-sized chemistry sets.

Overlooking the English Channel lies La Hague. Nestled amongst a mosaic of Normandy fields, it's the grandest nuclear recycling centre in the world. France pulls about 1,200 tonnes of spent fuel out of its reactors annually; running flat out, La Hague can churn through *2,000* tonnes in that time. It's a nuclear tour de force that represents most of the world's spent fuel sorting capacity.[2]

Inside La Hague, chemists stir spent fuel rods into boiling vats of nitric acid from behind leaded windows and concrete walls, as a sweet tooth stirs sugar into a steaming cup of tea. Solid lumps of high-level waste chemically disintegrate as they dissolve in the concoction.

The chemists infuse the mixture with a series of molecular adhesives. These adhesives are discerning. They preferentially stick to certain elements, whilst paying little heed to others. One by one, they're stirred into the high-level waste solution and, one by one, they're sorted into purified portions of uranium, plutonium, and fission fragments. Minor transuranics, like neptunium and americium, wind up with the fission fragments.

France mixes its fission fragments with molten glass, which cools and solidifies when it's poured into stainless steel canisters. The glass traps the radioactive atoms, priming them for eventual disposal in a geological repository. And as for the uranium and plutonium? Once sorted, they can be recycled.[3]

Recycled uranium

The major component of spent fuel, enriched uranium, can be turned back into fresh nuclear fuel rods. About 95 per cent of the spent fuel is recycled in this way.

France doesn't have indigenous reserves of uranium, and so it refabricated its own from recycled uranium. It's another example of the long-term nuclear mindset. This national asset will protect France should the price of uranium fluctuate. It's stashed away 34,000 tonnes so far, which contains as much energy as 490 million tonnes of coal or 2.3 *trillion* barrels of oil.[4]

Recycling uranium makes natural uranium reserves go further by yielding more megawatt-hours of electricity for every kilogram we dig from the ground, thus diluting the environmental impact – namely mining – of nuclear power even further. More than 70 nuclear reactors worldwide use (or have used) recycled uranium to generate power.[5]

Before we turn recycled uranium back into fuel for light-water reactors, though, we have to re-enrich it by giving it a whirl in the centrifuge. The fact it's enriched slightly to begin with gives us a head start on uranium that's fresh from the ground. But we can only recycle uranium so many times in light-water reactors. Occasionally, rather than splitting, a uranium-235 atom absorbs a neutron, thereby transforming into uranium-236. This is bad news for a nuclear reactor because uranium-236 is a neutron poison that severs links in chain reactions. When we chemically separate the uranium from spent fuel, we unavoidably separate all the isotopes together. And then when we re-enrich that uranium – to increase the proportion of fissile uranium-235 – we unavoidably enrich it in poisonous uranium-236 too. The fuel becomes slightly less potent with each round of recycling. For this reason, uranium is normally recycled only once before it's retired.[6]

Things are a little easier with heavy-water reactors. Heavy water is such a good neutron moderator that it's enough to send depleted uranium critical. In 2010, a heavy-water reactor in China generated electricity from the uranium sorted from second-hand light-water reactor fuel directly, without going through re-enrichment. Heavy-water and light-water reactors exist in symbiosis; the former can burn 'waste' from the latter.[7]

Fifty tonnes of recycled uranium from a light-water reactor could keep a heavy-water reactor going for about a year. That means France's stockpile could – if they burned it in, say, a CANDU reactor – power a nuclear reactor for *680 years*. All that energy, if the waste was buried underground, would be squandered.[8]

The Finnish model – from fuel, to energy, to waste, to disposal – approaches nuclear power with a *linear* mindset. But the French model – from fuel, to energy, to recycling, and then back to fuel – is a *circular* mindset.

The fissile rule

Nuclear fissility is an unusual property. Plenty of atoms will split if you wallop them hard enough with a neutron. But to split after being hit by a slow neutron – a neutron that's been moderated – is a characteristic possessed by just a few isotopes. These are the atoms that form the basis of nuclear power generation by sustaining fission chain reactions. Uranium-235 is the most famous fissile isotope. But are there others?

There's a rule of thumb for finding fissile isotopes, described first by nuclear scientist Yigal Ronen in 2006. The rule is as follows: for isotopes of elements between thorium (element number 90) and fermium (element number 100, named after Enrico Fermi), you triple the number of protons and then subtract the mass number; if your answer is either 41, 43, or 45, then it's fissile. It's weird, but it works.[9]

To illustrate this quirk in action, consider the two natural uranium isotopes: uranium-235 and uranium-238. Both have 92 protons. For uranium-235, the fissile rule (triple 92, minus 235) returns the number 41; uranium-235, therefore, ought to be fissile, which it is. For uranium-238, the rule (triple 92, minus 238) returns the number 38; uranium-238, therefore, ought not to be fissile, which is exactly the case.

This maths yields a list of 33 fissile isotopes. Each isotope could sustain fission chain reactions and generate electricity in principle. But it takes more than being fissile to turn turbines in practice. To start, 24 of those 33 fissile isotopes have ruinously short half-lives. The shorter an isotope's half-life, the more radioactive, and so the more troublesome it is to handle in large quantities and safely sculpt into fuel rods. Most of them, like americium-244, with its 10-hour half-life, also decay out of existence far quicker than we could ever load them into a nuclear reactor.[10]

And then there's the small matter of availability. Of the nine fissile isotopes with longer half-lives, only one – uranium-235 – exists naturally. To use any of the others to power a nuclear reactor, we must synthesise them first. Coaxing millionth-of-a-gram specks of these isotopes through experiments is one thing. Churning out enough to power hundreds or thousands of reactors is something else entirely.

To generate nuclear power using atoms besides uranium-235, therefore, we need fissile isotopes with relatively long half-lives that we can forge in enormous quantities. Whittling down the list of 33, only 2 more meet these criteria: plutonium-239 and uranium-233.

Nuclear squander

The most abundant plutonium isotope in spent fuel is fissile plutonium-239. This means we can recycle plutonium – a by-product of generating nuclear power – *and use it to generate more nuclear power*. Recycling plutonium alongside uranium ties another loop in the nuclear power cycle.

A *circular* fuel cycle – turning by-products into energy – is unique to nuclear power. Fossil fuels are *linear* because we only get one shot at extracting their energy: once those carbon and hydrogen atoms chemically acquaint themselves with oxygen, we can't burn them again. Worse than that, the by-product – carbon dioxide – is lost to the atmosphere, where it causes climate change.

At 141 tonnes, the civil stockpile of plutonium at Sellafield in the UK is the largest in the world. Whilst it's weighty, its high density means it would fit into a cube just 2.3 metres across. But it would take at least 100,000 years for its radioactivity to decay back to that of uranium ore if left to its own devices. Something must be done in the meantime.[11]

The UK *could* recycle its plutonium and use it to generate electricity. There's enough plutonium at Sellafield to power the pair of new reactors at Hinkley Point C – which together will generate enough electricity to satisfy 3 million people annually – until the second decade of the next century. We wouldn't need to excavate a speck of uranium from the Earth to generate that energy. It's already there, waiting.[12]

But the UK classifies the prodigious plutonium pile as a 'zero value asset'. That sounds a lot like a euphemism for 'waste'. And after decades of agonising over what to do with it, the British government announced in 2024 that it won't be using the plutonium to power its new fleet of reactors. That leaves one option: burying it in Onkalo-style catacombs. All that latent emissions-free energy will be put out of reach.[13] Burying plutonium isn't yet government policy, but there isn't another long-term option. The UK isn't alone. Most nuclear nations won't be recycling their plutonium, either. The political appetite is diminished by anxieties over cost escalations, fears of security breaches, and the effort of building the industrial infrastructure required to recycle spent fuel.

Britain amassed its stockpile of plutonium by chemically sorting it from spent fuel, in much the same way as at La Hague. Motivated by fears of uranium shortages, it intended to turn the plutonium into electricity. But the discovery of enormous global uranium reserves, and the global nuclear wind-down in the wake of Chernobyl, prevailed against those good intentions. Its nuclear energy cycle has been linear and open-ended ever since, and it amassed a plutonic albatross. France, on the other hand, tied a loop in its nuclear power cycle and has been using plutonium to make electricity since 1987.[14]

Water-moderated reactors don't run on pure plutonium, and so we blend it with uranium to make 'mixed oxide fuel', or *MOX* for short. Plutonium is so fissile that even if it makes up just a few per cent of the fuel, it will send natural and even *depleted* uranium critical in a light-water reactor. It eliminates the need to re-enrich the uranium once it's chemically separated. It also means we can turn those hex tails – the 'waste' uranium left over after enrichment – into useable fuel.[15]

France currently fabricates almost all the MOX in the world and uses it to generate 10 per cent of its nuclear electricity, which means its electric cars are – at least partly and indirectly – run on plutonium. (DeLorean time machines remain a pipedream.)[16]

With decades of learning by doing, France has made nuclear waste recycling economical. It got 40 per cent cheaper per kilowatt-hour of electricity in the past decade alone. And whilst recycling nuclear fuel *does* make nuclear power pricier, it's by a pittance: in France, it adds less than €1 per month to a typical household electricity bill. Most of the cost of nuclear electricity is still incurred in the up-front cost of building power stations.[17] Recycling nuclear waste is like recycling plastic bottles: we accept some inconvenience and a small economic cost, but in return we build a system that's more sustainable.

Recycling high-level waste by turning it into MOX squeezes about 30 per cent more energy from the uranium we dig from the ground.[18] The less uranium we need to mine, the better, because excavating uranium ore from the Earth is the most environmentally detrimental aspect of nuclear power generation. If the world turns to nuclear power over the coming decades, our demand for uranium will increase commensurately. It is our responsibility to make a hard-won natural resource go further.

Incinerating plutonium in nuclear reactors has another obvious advantage: *it gets rid of the plutonium*. France generated about 350 per cent more nuclear energy than the UK over the past 6 decades, but its plutonium stockpile is 25 per cent smaller.[19]

Burning plutonium doesn't make those atoms vanish, though. It just splits them into fission fragments. But whereas plutonium remains radioactive for hundreds of millennia, fission fragments remain radioactive for a fraction of that time. By turning plutonium into fission fragments, we shorten the radioactive lifespan of spent fuel. This eliminates the need to store nuclear waste underground for *geological* timescales; we just need to store it for *historical* periods. Long-term storage is still required, but it's a much easier exercise when the catacombs only have to last a thousand years.

'Nuclear waste' is inaccurately named, so I'm proposing a swift, orderly change to 'nuclear squander'. If we bury it, our descendants may well end up mining it and using it to power their reactors.

⚛ ⚛ ⚛

So far, we've considered our relationship with energy through the looking glass of today's problems: promoting prosperity through industrialisation and energy abundance; maintaining the living standards of the developed world; protecting the natural environment and mitigating climate change by phasing out fossil fuels. These are problems that nuclear power can solve in the coming century.

But what lies beyond? What comes after net zero?

Fissioning, fast and slow

When atoms split, neutron sparks abound. But newborn neutrons travel at breakneck speed. These so-called *fast neutrons* are unlikely to form links in chain reactions; they tend to sail right past fissile atoms, which fizzles the reactor towards a sub-critical whimper.

Slow neutrons (called 'thermal' neutrons in nuclear parlance), as Fermi taught us, are far more likely to interact with uranium nuclei and thus tilt a reactor towards criticality. That's why **all** conventional reactors – sometimes called *slow reactors* (or *thermal* reactors) – use moderators: graphite in Fermi's case; water in 95 per cent of the present-day global fleet.

None of this is to say fast neutrons *can't* induce nuclear fission. They can, but they're just unlikely to do so.

You can't defy the laws of probability, but you *can* game them. By increasing the number of fissile atoms inside a nuclear reactor we tip the balance towards criticality. This is the rationale behind uranium enrichment in slow reactors. But if we increase the number of fissile atoms enough, even fast neutrons can send a reactor critical.

Normal fuel isn't potent enough to go critical with fast-neutrons because it's only 5 per cent uranium-235, tops. Sending a reactor critical using fast neutrons requires something with a little more kick . . .

Enriching uranium far beyond 5 per cent is one way to do it. By sending it merrily around the centrifuges again and again, we can – and sometimes do – enrich it all the way up to 100 per cent. About 20 per cent enrichment does the trick for fast neutrons.[20]

But plutonium is superior to enriched uranium. All four of its common isotopes – plutonium-239, plutonium-240, plutonium-241, and plutonium-242 – have better odds of splitting after being struck by a fast neutron than uranium-235. That means it's easier to send critical. And, compared to uranium-235, plutonium atoms release more neutrons when they fission. That makes their odds of sustaining a chain reaction even better. Plutonium is also common – there's no shortage of it in nuclear 'waste'.[21]

By using these super-fuels, we eliminate the need for a moderator, and can sustain fission using fast neutrons. And therein lies the essence of so-called *fast reactors*. Only about two dozen of these reactors have ever existed, and only nine have operated commercially, but they've collectively racked up more than 400 years of runtime.[22]

So, why bother with fast reactors? Why not stick with slow reactors, like the water-moderated ones that dominates nuclear power today? There are two main reasons.

The first is that they burn transuranic elements that would otherwise become nuclear waste. Fast neutrons are far more likely than slow neutrons to split transuranic atoms. We can therefore chemically sort them from high-level waste, load them into fast reactors, and harvest energy from them. By breaking them into relatively short-lived fission fragments, we bypass the geological deep time required to decay them into nothingness. Fast reactors, in effect, incinerate long-lived nuclear waste.[23]

The second reason is that they could supply energy until the end of this millennium.

Breeder reactors

It can sometimes feel like we're trying to squeeze energy from a stone. Only 0.7 per cent of uranium atoms in nature are fissile uranium-235. Even if we burned every last one through repeated recycling, we'd *still* only put 0.7 per cent of the uranium atoms we dig from the ground to good use. That means in slow reactors, more than 99 per cent of them are ultimately squandered. True, we create fissile plutonium inside slow reactors when neutrons fertilise uranium-238, but it's a happy accident that makes the original uranium go about a third further. That's an effective increase from 0.7 per cent 'useful' to about 0.9 per cent 'useful'. It's better than nothing, and gets rid of the plutonium, but it's a marginal gain.[24]

And it's with this accident of nuclear physics that one of the most radical ideas in the history of nuclear power was born: making our uranium go much further by *intentionally* creating plutonium inside fast reactors.

Fast neutrons generate more neutron sparks than slow neutrons.[25]

If slow reactors conjure a drizzle, fast reactors summon a tempest. There are so many neutrons that — even after chain reactions have consumed some of them — plenty go spare. We can put those surplus neutrons to good use. By intentionally directing them towards fertile uranium-238, we can cultivate large amounts of fissile plutonium-239. If we create fissile atoms faster than we burn them, the number of fissile atoms in our reactor *increases* over time. Inside fast reactors, *we can create more nuclear fuel than we burn*.

Fermi floated the idea back in 1944, less than 18 months after he'd sent the world's first reactor critical. Minutes from a meeting between him and several other Manhattan Project scientists — since declassified — record its inception:

> If all the neutrons are captured, the overall balance would be that for every atom of 49 destroyed, two atoms of 49 would be produced.[26]

'49' was the code name for plutonium-239, enciphered by the final digits of its position on the periodic table (94) and its mass number (239). In other words: if we got it right, we'd harvest energy *and* end up with more fuel than we started with. When a fast reactor creates more fissile atoms than it consumes, we call it a *breeder reactor*.

If we breed fissile plutonium-239 from fertile uranium-238, we can recycle it back into nuclear fuel. Then we can use *that* nuclear fuel to breed even more plutonium-239, and then recycle *it* back into more nuclear fuel. And so on. By closing the circle on nuclear power we could — in principle — use every last uranium atom we dig from the ground. It almost eliminates the need for uranium mining. A breeder reactor generates power *and* fuel.

In practice, no nuclear cycle could ever be perfect enough to use every last atom, but breeders still make far more efficient use of uranium. Imagine you have 6.5 grams of uranium, about the same mass as a grape. If you burned the uranic grape in a slow reactor, you'd generate a person's worth of electricity for a fortnight. If you burned it and recycled it back into French-style MOX fuel, you'd squeeze an extra 4 days' worth of electricity from it. Not bad. But

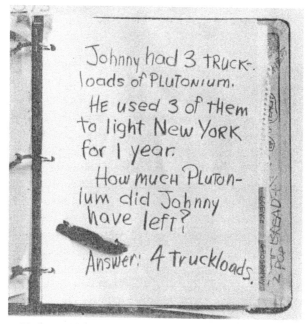

An old joke, scribbled into a notebook, captures the essence of breeder reactors. Image courtesy of United States Department of Energy Office of Scientific and Technical Information, circa 1950s.

if you burned it in a fast reactor, you'd have power for more than *2 years*. Multiplied over somebody's 80-year lifetime, that's uranium weighing no more than a small bunch of grapes. Using a comparable mass of fossil fuel, we wouldn't even generate enough electricity to last an hour.[27]

The United Kingdom's 141-tonne stockpile of 'waste' plutonium could power somewhere between 6 and 10 breeder reactors, each capable of churning out 1,000 megawatts of electricity. That's at least a doubling of the UK's current nuclear capacity *by burning nuclear waste*. And it wouldn't run out of plutonium fuel for many centuries because it could breed more from its 110,000 tonne stockpile of 'waste' depleted and low-enriched uranium. It would be nuclear power almost for eternity.[28]

EBR-1

Breeder reactors were a tantalising prospect in the mid-twentieth century when uranium scarcity was on everybody's mind. If the world ran out, it would spell nightfall for the Atomic Age.

In 1945, Fermi – with characteristic vision – said, 'The country which first develops a breeder reactor will have a great competitive advantage in atomic energy.' Two years later, the nascent United States Atomic Energy Commission approved plans for a prototype to be built in the vast Idaho Desert. With a mid-twentieth century let's-get-things-done attitude, the imaginatively named *Experimental Breeder Reactor-1* reached criticality within 4 years. It was the first nuclear reactor in the world to breed plutonium *and* generate electricity at the same time.[29]

Experimental Breeder Reactor-1 was beautiful. Scientists filled its core with pencil-thin fuel rods fashioned from highly enriched uranium (94 per cent uranium-235). Like bees filling comb with honey, they arranged all 217 of them in a tessellating hexagon pattern. The honeycomb core was about 20 centimetres wide and 22 centimetres tall.[30]

Scientists couldn't use water to stop the core overheating as they would in a conventional reactor; water is a moderator that slows down neutrons, and the whole point of a breeder reactor is to keep the neutrons travelling fast. They needed a coolant that neither moderated nor poisoned the chain reaction. An alloy of sodium (Na) and potassium (K) nicknamed 'knack' – a metal that's liquid at room temperature – fit the bill. The liquid metal, shimmering as it flowed, flooded the core of the reactor and drew nuclear heat into an exchanger.[31]

To catch the neutrons that missed uranium-235 atoms, scientists encased the hexagonal core in a 'blanket' of natural uranium. The blanket took the form of rods and bricks, and being natural – 99.3 per cent uranium-238 – it contained plenty of fertile seeds to capture stray neutrons.[32]

And it *worked*. Six months after the reactor went critical, scientists obtained the first proof of breeding. Chemical analysis of the fuel and blanket revealed that for every fissile atom of uranium-235 burned, one fissile atom of plutonium-239 had been created. Experimental Breeder Reactor-1 had transformed inert uranium-238 — practically useless in slow reactors — into nuclear fuel.[33]

The whole thing was just 80 centimetres across. Here's a sketch of it from a 1950s textbook:

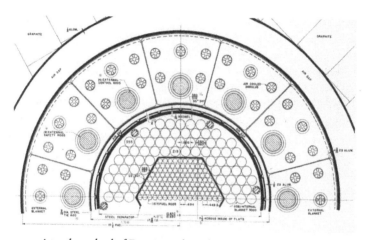

A top-down sketch of Experimental Breeder Reactor-1. All dimensions are in inches. The reactor wraps around full-circle, but it's been cut away in this sketch. Image courtesy of Argonne National Laboratory.

During an experiment in 1955, the power of the core jumped unexpectedly from 50 watts — enough to power five lightbulbs — to 1 *million* watts — enough to power 100,000 lightbulbs — in about 3 seconds. It was the *Experimental* Breeder Reactor, after all. In a panic to scram the reactor, an operator accidentally hit the slow-moving control rod button instead of the rapid emergency shutdown button. The core's temperature soared, and the fuel rods partially melted. But nobody was hurt, and nobody was exposed to radiation. Engineers replaced the damaged core with a new one that ran until 1963, at which point the reactor was switched off for good. Today, the Experimental Breeder Reactor-1 building is a public museum. I'd love to visit someday — it's on my nuclear reactor bucket list.[34]

The third fuel

When we turned the handle on the fissile rule, we found three chain-reacting atoms that could usefully power nuclear reactors. We can dig uranium-235 from the ground, and we can alchemise plutonium-239 inside breeder reactors. But what about uranium-233?

Move two tiles backwards from uranium on the periodic table and you'll find an element named after the hammer-wielding Norse god of thunder and lightning: *thorium*, element number 90. Thorium is idiosyncratic because almost all of it is just one isotope; if you counted out 10,000 thorium atoms, all but two would be thorium-232. And with a 14 billion-year half-life, thorium-232 is barely radioactive. Despite being in the elemental vicinity of uranium and plutonium, thorium cannot sustain chain reactions. It is not fissile and so we can't use it as nuclear fuel. But thorium *is* fertile.[35]

When a thorium-232 atom absorbs a fast neutron, it grows to become thorium-233. Less stable than its lighter doppelgangers, the thorium-233 atom beta decays swiftly, with a half-life of 22 minutes. It becomes protactinium-233. Then protactinium-233 beta decays, with a 27-day half-life, into an unfamiliar isotope of a now-familiar element — uranium-233.

Uranium-233 *is* fissile. And its 160,000-year half-life means it sticks around plenty long enough for us to fashion fuel rods from it. Uranium-233, grown from thorium seeds, joins uranium-235 and plutonium as the third fissile fuel. It gives us another type of atom from which we can sustain chain reactions and harvest emissions-free energy.

Uranium-233 doesn't exist naturally, but we can breed it from thorium by analogy to breeding plutonium from uranium. Therein lies the rationale behind *thorium reactors*. Instead of surrounding a breeder reactor with a blanket of fertile uranium, in thorium reactors

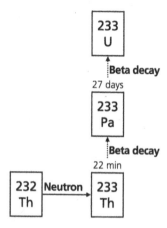

we use a blanket of fertile – you guessed it – thorium. No nation has grander plans for this than India.

Homi Bhabha was born in 1909 in Mumbai. He hardly slept as a child, which worried his parents. They sought medical advice from a French doctor, who, upon meeting Bhabha, realised there was nothing wrong with him – he was just hyperactive, with a mind that whirred relentlessly. It's a trait that characterised Bhabha throughout his life. Few individuals have singularly influenced the trajectory of a nation's nuclear masterplan more.[36] As an adult, he said:

> What comes after death no one knows. Nor do I care. Since, therefore, I cannot increase the content of life by increasing its duration, I will increase it by increasing its intensity.[37]

At the age of 18, Bhabha travelled to England to study at Cambridge University. His father and uncle expected him to graduate as an engineer and work in the steel industry, but Bhabha had other ideas. In 1928, he said in a letter to his father:

> For, each man can do best and excel in only that thing of which he is passionately fond, in which . . . he is in fact born and destined to do . . . I am burning with a desire to do physics.[38]

Seven years after arriving in Cambridge, he'd earned his PhD in theoretical physics by researching the science of radiation from outer space.

In 1939, whilst Bhabha was holidaying back in India, World War II broke out in Europe. His plans to return to Cambridge and continue his research were thwarted. He took up a post as Reader in Theoretical Physics at the Indian Institute of Science. Three years later, he was made professor, and he decided to stay in his homeland to fulfil his 'duty to stay in one's country and build up schools comparable with those in other lands'.[39]

Determined to establish a school dedicated to physics research, Bhabha established the Tata Institute of Fundamental Research in 1945. Bhabha welcomed bright, talented scientists from across India, and let them pursue their scientific interests with freedom. 'A scientific institution, be it a laboratory or an academy, has to be grown with great care like a tree,' he said.[40]

Meanwhile, Bhabha watched the revolution in nuclear science unfolding in the West. The idea of a society powered by the atom gripped him. In early 1948, he penned a letter to Indian Prime Minister Jawaharlal Nehru urging him to lay the foundations of a domestic nuclear programme. He told Nehru 'within the next couple of decades atomic energy would play an important part in the economy and industry of countries'. A visionary, Bhabha wrote with a tone of foreboding: 'if India did not wish to fail even further behind the industrially advanced countries of the world, it would be necessary to take more energetic measures to develop this branch of science.'[41] It was the same ideal that lay at the core of 'Atoms for Peace', but five years before President Eisenhower's famous speech. Nehru took Bhabha's words to heart. The Atomic Energy Commission was founded in India four months later.

Nuclear power could power India as it grew into a developed nation. In 1954, Nehru declared to his parliament '. . . the use of atomic energy for peaceful purposes is far more important for a country like India, whose power resources are limited, than for a country like France, an industrially advanced country . . . Atomic energy is a tremendous tool for the benefit of humanity, whether it is disease or poverty.' The Atomic Energy Establishment Trombay was thus established to fulfil this ideal.[42]

Bhabha wanted India to be self-sufficient and capable of producing homemade nuclear energy. But he had a problem. India doesn't have much indigenous uranium. It has enormous reserves of another element, however: thorium. To turn thorium into fuel, India needed a breeder reactor. At the time, it didn't have a single nuclear power station, and there were no commercial breeder reactors anywhere on the planet. Experimental Breeder Reactor-1 was 3 years – and half the globe – away.

But Bhabha was thinking for the long term. Epic nuclear programmes take time. In contrast with today's climate of short-sightedness, he imagined India's nuclear programme evolving through a series of three phases over decades and centuries. Each phase would use the types of reactors we've encountered already. I'll let Bhabha describe the first phase himself:

> The first generation of atomic power stations – based on natural uranium – can only be used to start off an atomic power programme . . .[43]

So, a fleet of nuclear reactors fuelled by natural – i.e. not enriched – uranium: pressurised water reactors, moderated by heavy water. India is blessed with hydro-enabling mountainous terrain, which could supply all the electricity it needed to concoct homemade heavy water. That's why, to this day, India has the largest fleet of heavy-water reactors outside of Canada.[44]

Burning natural uranium to generate electricity would kick-start India's nuclear programme. But the fissile by-product of spent fuel – plutonium – would enable the second phase of Bhabha's plan. Eventually, there would be so much plutonium in the nuclear 'waste' that it could be chemically sorted and recycled to make new fuel. It's far easier to chemically sort *old* spent fuel than *new* spent fuel, because it's much less radioactive.

> The plutonium – produced by the first generation power stations – can be used in a second generation of power stations designed to

produce electric power and convert thorium into uranium-233 (or depleted uranium into more plutonium with breeding again) . . .[45]

So, a fleet of nuclear reactors that burn recycled plutonium. If they were breeder reactors, those cores could be surrounded by blankets of depleted uranium – the waste left over from the heavy-water reactors in phase I – to breed even more plutonium fuel. They'd essentially be self-sustaining.

But the phase II reactors were just stepping stones. The real purpose was to surround their cores with blankets of India's greatest nuclear asset: thorium. By turning thorium into uranium-233 India would, eventually, amass the fuel to power the phase III reactors:

> The second generation of power stations may be regarded as an intermediate step for the breeder power stations of the third generation, all of which would produce more uranium-233 than they burn in the course of producing power.[46]

So, a fleet of nuclear power stations that burn recycled uranium-233 *and make more uranium-233 as they go*. Thus, the plan reaches its zenith, and India churns out massive amounts of electricity, all whilst replenishing its own fuel. *Ad infinitum*.

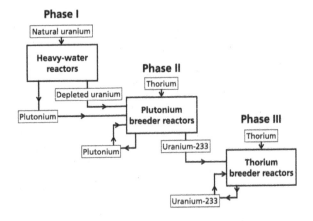

Phase I: heavy-water reactors powered by natural uranium. Phase II: breeder reactors, powered by recycled plutonium, with depleted uranium and thorium blankets. Phase III: breeder reactors, powered by recycled uranium-233, with thorium blankets. The reactors of phase III burn *and* breed uranium-233 in a closed loop:

Bhabha's vision represents the true closure of the nuclear power cycle. Every last atom dug from the ground is put to good use. Mining essentially stops, and nuclear power's environmental footprint is reduced to practically nothing. He presented his vision in 1954, and it was adopted officially by the Indian government 4 years later.[47]

In 1966, Bhabha was flying to Vienna to attend a meeting at the International Atomic Energy Agency. The aeroplane crashed into the side of Mont Blanc in the Alps. There were no survivors. A year later, the Atomic Energy Establishment Trombay was renamed the Bhabha Atomic Research Centre in his honour, and today, he's remembered today as the father of the Indian nuclear energy programme.[48]

Boom and bust

In the decades following Experimental Breeder Reactor-1, Japan, India, the Soviet Union, Germany, France, and the UK embarked on ambitious research programmes to develop their own breeder reactors. In test reactor after test reactor, the technology was proven, and the full energy potential of uranium was unlocked.[49]

By 1971, the Nobel Prize-winning nuclear chemist who discovered plutonium, Glenn Seaborg,* said, 'Our relationship to all our basic needs – food, water, shelter, clothing, a liveable environment – could change drastically' because of breeder reactors. Back then, they were projected to represent a third of the world's nuclear power capacity by the year 2000.[50]

---

* Seaborg led the group that discovered plutonium in 1941 in Berkley, California, but they kept it secret until after World War II, because plutonium-239, like uranium-235, can be turned into atomic bombs. Seaborg shared the Nobel Prize in Chemistry 1951 with McMillan. Element number 106 is named *seaborgium* ('Sg') in his honour.

But in the end, they comprised a paltry 0.2 per cent. Only two operated commercially: France's prototype Phénix reactor and Russia's Beloyarsk-3 reactor.[51]

Natural uranium turned out to be more plentiful than initially thought. And then, after the second wave of the nuclear rollout was broken by Chernobyl in the late 1980s, the global expansion of nuclear power halted. Uranium supplies soared as demand plummeted. Pressure on uranium supplies abated, and the urgency to develop breeder reactors evaporated.[52]

Today, a few nations are dabbling in breeder reactor development, namely India, Russia, and China. But there's no broad, concerted effort internationally. Only two operate commercially – Russia's Beloyarsk-3 and Beloyarsk-4 reactors – and conventional slow reactors outnumber them by more than 200 to 1.[53]

They're pricier than slow reactors and can't compete financially. It's a nuclear catch-22: breeders won't become cheaper until we start building them in large numbers, but we won't build them in large numbers until they become cheaper. Advancing them to the point where they can supply electricity and heat to society requires extensive research and development, an end made possible only by government backing and international collaboration.

※ ※ ※

Bhabha's dream has not yet come to fruition. India today has a modest fleet of heavy-water reactors, but it hasn't transitioned to phase II. It *did* use recycled plutonium to power a prototype breeder reactor in the mid-1980s, but it was beset by engineering challenges and outages. Today, nuclear power stations generate less than 3 per cent of the nation's electricity.[54]

But Bhabha's dream isn't dead. A closed nuclear power cycle based on breeder reactors – with Finnish-style disposal of the fission fragments – is the apex sustainable energy system. More than 95 per cent of the atoms in spent nuclear fuel are either fissile or fertile. By disposing of it wholesale, we're throwing the baby out with the bath water. (Or, rather, we're throwing the emissions-free energy out with the fission products.) If we used every atom we dug from the ground, nuclear's environmental impact would fade into insignificance.

Until recently, breeder reactors were academic curiosities that never truly made it beyond the prototype phase. Net zero has stoked interest once more. In 2022, the International Atomic Energy Agency – drawing on experience from nine nations spanning four continents – concluded there are 'no fundamental insurmountable barriers' that would stop the mass deployment of thorium reactors. In March 2024, India loaded fuel into the core of its first commercial breeder reactor – the *Prototype Fast Breeder Reactor* – which is set to go critical imminently. And just a few months later, Bill Gates' TerraPower began constructing its first *Natrium* reactor – a 350-megawatt breeder reactor that uses uranium as fuel and liquid sodium as a coolant – in Wyoming, close to a retiring coal-fired power station.[55]

Conventional reactors – powered by slow neutrons – could meet humanity's energy needs in the short and medium terms. But breeder reactors are a solution for the long term. Whilst rolling out a massive fleet of pressurised water reactors over the coming decades, we *must* build on the research and development started in the mid-twentieth century. Whilst breeder reactors aren't *technically* renewable – nuclear fuel doesn't exist in infinite amounts – they are essentially renewable on human timescales. Indeed, a seminal report on sustainability by the United Nations in 1987 includes breeder reactors in a list of renewable energy sources. We must commercialise breeder technology; net zero compels us to.[56]

So, for how long could we cleanly power the entire world with nuclear?

There are about 8 million tonnes of mineable uranium in the ground. (That's just the stuff we know about; we discover more and more each year, but let's stick with that number for now.) Let's say we stopped using *all* fossil fuels for *all* energy purposes and replaced them with nuclear, tomorrow. We could run the world for about 8 years using a once-through, linear nuclear power cycle. If we tied a loop in that cycle and recycled the plutonium as French-style MOX, we'd have energy for a decade or so. But if we burned it in fast reactors, the uranium would last about 440 years.

And then there's the world's stockpile of depleted uranium 'waste', left over as hex tails after enrichment. It exceeds a million tonnes. Burning it in breeder reactors would give us almost 7 decades of energy. And *then* there's the 264,000 tonnes of spent fuel that hasn't yet been recycled; we could burn that in breeder reactors, too, which would give us another 14 years *and get rid of all the long-lived nuclear waste*.

And finally, there's the third fuel. Known thorium reserves stand at 6.2 million tonnes, which would give us another 340 years of power.[57]

In all, there's enough to last us almost 900 years. These are, clearly, back-of-the-envelope calculations. Expanded renewable generation didn't feature in my sums, nor did future discoveries of new uranium and thorium reserves. But the rough numbers illustrate an important point: we could power the world cleanly for most of the third millennium with nuclear. By then, maybe we'll have cracked fusion.

Beyond the uranium in the ground, there are 4.5 billion tonnes dissolved in the oceans. It could power the world for the next *quarter of a million years*. That is longer than our species has walked the Earth. The concentration of uranium in ocean water is tiny – a scant 150 trillionths of a gram in a drop of water, equivalent to a salt grain stirred into a bathtub – which makes it difficult to extract. But if we found a way to harvest it economically, we'd tap a source of uranium that – for all intents and purposes – is inexhaustible. Research and development is ongoing with that end in mind. We've got centuries to get it working, anyway.[58]

Devising a truly sustainable energy system – such as the one envisioned by Bhabha – requires us to think far beyond net zero. Establishing a closed nuclear fuel cycle based on breeder reactors is how we embed the wellbeing of future generations and the planet into our actions today.

Chapter 8. Radiophobia

Nothing in life is to be feared, it is only to be understood.

– *Marie Curie*, date unknown

I'm a nuclear chemist. Radiation equivalent to 1.1 trillion gamma rays strikes my body each passing year. Those gamma rays – 36,000 of them every second on average – zap my cells as they crash silently through my tissues and bones. Some of those cells perish. Others survive unscathed. But some may see their genetic code scrambled, causing them, over months and years, to mutate and multiply into a rampant cancer.

It sounds pretty bad, doesn't it? But before you lament my misfortune, consider this: everything in the previous paragraph would be true if instead of the words 'nuclear scientist', I'd written 'butcher', or 'baker', or 'candlestick maker'. That's because every human being spends their lives bathed in radiation. That includes you, right now, as you read these words, and it includes me, right now, as I write them. It even includes my cat who's watching me write this paragraph. Biology unfolds against a background of radioactivity. It's everywhere, for all of us, all the time.

More than 99 per cent of this background radiation arises from natural sources. The rest stems from new sources of radiation we've put into the environment.[1] There are also human activities that give us an extra dose, such as taking X-rays of broken bones and aeroplane flights. People are seldom exposed to even mildly dangerous levels, and yet the trefoil symbol – a central atom with three blades representing alpha, beta, and gamma radiation – is a symbol of *radiophobia*: the irrational fear of radiation.

In films it signals unseen peril; it's appropriated by anti-nuclear campaigners as they sow the seeds of disinformation; it's mingled with the mushroom cloud's dark shadow. Over and over again, radiophobia manipulates and reinforces our perception that radiation is bad news.

Radiophobia is a spectre that conjures visions of cancers, birth defects, and, at least on the silver screen, glowing monsters. But how worried should we be? To answer that question, first we need to make the distinction between *radiation* and *ionising radiation*.

Radiation encompasses all light and sound. It's simply the motion of energy in the form of particles and waves: electromagnetic waves in the case of light, and vibrational waves in the case of sound. The microwaves that warm your dinner and radio waves that play your music are radiation; so are Wi-Fi and phone signals.

Ionising radiation is waves and particles that carry enough energy to knock electrons from atoms. It sparks tiny chemical reactions in its target. Those chemical reactions have the potential to wreak havoc inside cells and tissues. It includes forms we're already familiar with, such as fragments of atomic nuclei and electrons. It also includes gamma rays and their less energetic counterparts, X-rays.

'Radiation' is often used interchangeably with 'ionising radiation', even in the scientific literature. And *ionisingradiophobia* doesn't have the same ring to it. With that in mind, for the sake of brevity I'll drop the word 'ionising' from herein.

Counting in sieverts

We count our everyday experience of the world with an assortment of units. We measure journeys in *miles* and *kilometres*; we wait *minutes* or *hours* for the next train; we pour wine in millilitres, and we pull pints in . . . pints. To quantify the health effects of radiation exposure, we're obliged to add a new unit to our metrological repertoire: the *sievert*. (A vintage unit – still used widely in the USA, a nation with a fondness for anachronistic units – is the *röntgen equivalent man*, or 'rem' for short; it's equal to 0.01 sieverts.)

One sievert is the damage to the body caused by about 450 trillion gamma rays. The sievert accounts for the health effects of different types of radiation; alpha particles and neutrons do more damage than the same number of beta particles and gamma rays.[2] It also accounts for which parts of the body are exposed; 450 trillion gamma rays to the fingertips are far less consequential than the same number to the chest, for example. Sieverts allow us compare radiation doses, like for like.

It's quite easy to be exposed to beta and gamma radiation. All you have to do is stand next to the radioactive object. Nimble beta particles and zippy gamma rays pass through the air and into your tissue with ease. Many of them will zap your cells, though most will stream through your body as if it wasn't there. Bulky alpha particles, on the other hand, come to a halt quickly. They're blocked completely by a finger's width of air or a fine layer of cotton. I could hold an apple-sized lump of pure plutonium-239 – an intense alpha-emitter – in the palm of my hand, and my radiation dose would be zero because my outermost layer of skin would block it. Alpha radiation can't do damage from the *outside* of your body, but it can do damage from the *inside*. To receive a dose from an alpha-emitter, you have to either inhale it, inject it, or eat it.

Radiation is invisible and so you cannot perceive a sievert directly. It has neither sound nor taste nor smell nor texture. Its invisibility doubtlessly enhances its supposed insidiousness; we tend to fear what we can't see. To count it, therefore, we must extend our senses with a toolkit of radiation detectors.

My labmates and I wear our very own personal radiation detectors at work. We call them *dosimeters*. The detectors – pinned to our chests like name badges – go wherever we do. They tot up the sieverts as we busy ourselves with mass spectrometry and the like. Every 3 months, we send them to the Sellafield dosimetry centre, where health physicists log our dose in our radiation records.

If you wore a dosimeter 24/7 for a year and then sent it off to your local health physicist, you'd be in for an anti-climax. Your dose wouldn't be anywhere close to even a single sievert. A sievert is more radiation than most people are exposed to over their entire lives. Background radiation varies from nation to nation, and even within nations, but the global average over a year is a meagre 0.0024 sieverts. We normally measure sieverts in their thousandths, therefore, in which case the yearly average becomes 2.4 *milli*sieverts.[3]

It's only natural

If you're concerned about the 2 million or so gamma rays' worth of background radiation that crashed through your body in the past minute, I have a few suggestions for how you can lower your dose.

Wielding an umbrella made from lead would be a good place to start. The Sun is like a fountain from which tides of charged particles surge. Night-time is no escape; a storm of charged particles blows perpetually from distant stars and saturates the solar system in even more extraterrestrial radiation. If you could see cosmic rays beating down from outer space, it would be a ceaseless monsoon. Your lead umbrella would shield you from the celestial downpour, reducing your annual dose by 0.4 millisieverts, or 16 per cent.[4]

Cosmic radiation varies with altitude. As you climb higher through Earth's atmosphere, the protective air above you thins. Exposure soars 24-fold between sea level and a cruising altitude of 35,000 feet.[5] So avoid flying, too.

Next, you could line the floors in your house with lead sheets. They'd block the radiation that streams naturally from the rocks beneath your feet. If you've sheets to spare, use them to line your walls, too; they'd impede the radiation that emanates from all bricks, stones, and tiles. Or better still, demolish your house and build a new one from wood, but don't forget to lead-line your floor. For your inconvenience, you'd reduce your annual dose by 0.5 millisieverts, or 20 per cent.[6]

To reduce your dose even further, stop drinking. A typical glass of tap water contains a few hundred billionths of a gram of uranium, but can contain thousands of times more, depending on the geology of your local bedrock. That uranium fires out alpha particles as it works its way through your body. Once you're suitably dehydrated, stop eating, too. Potassium – of which a tiny dash is radioactive potassium-40 – exists in all foodstuffs; it propagates upwards through the food chain when plants draw it from the soil. Potassic foods, like potatoes and bananas, are particularly radioactive. Your body can't tell the difference between radioactive and non-radioactive potassium, and so you become radioactive when your body indiscriminately absorbs the nutrition. If you want to go the extra mile, abstain from eating Brazil nuts; their extensive root systems scavenge tiny amounts of radium from the soil, an alpha-emitting element so radioactive it glows in the dark when assembled in sufficient quantities. For your indefinite fasting, you'd reduce your annual dose by 0.3 millisieverts, or 12 per cent.[7]

And to round off your heroic efforts, stop breathing. The air hangs heavy with radioactive radon-222, an intermediate isotope in the uranium-238 decay chain. Radon gas seeps from the ground through fractures in the bedrock. Every time you draw breath, you flood your lungs with radon. It spits forth high-energy alpha particles with a half-life of 4 days, decaying through a further seven radioactive progenies as it works its way down the periodic table toward non-radioactive lead. The cessation of breathing would reduce your radiation dose by about 1.2 millisieverts, cutting your annual dose in half.[8]

All this depends on where on Earth you live.

Radon is by far the largest and most variable source of exposure. Its prevalence depends on the local geology – the more parental uranium in the bedrock, the more radon progeny seeps from it. It varies massively within countries, too. The uranium-rich granite in the south-west of the UK, for instance, gives residents of Cornwall an annual dose of 6.9 millisieverts; that's 2.5 times the national average, just from radon.[10] Holding your breath is the single biggest action you can take to reduce your dose to background radiation. Don't forget to exhale before you hold your breath, though; you

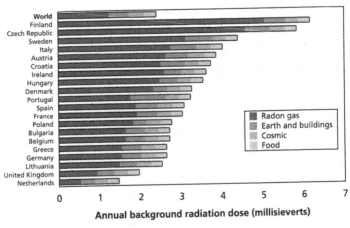

Background radiation from natural sources in twenty of the most populous European nations. I omitted anthropogenic sources of radioactivity, such as atom bombs and Chernobyl fallout, because they're imperceptible at this scale.[9]

need to purge your lungs of radon first. Or, if you want to get creative, breathe exclusively from bottled air; after 5.5 weeks, more than 99.9 per cent of the radon in the bottle will have decayed into stable lead.

Now, what you could do instead of my mischievous and ridiculous suggestions is nothing. Forget about it. Ambient background radiation is inescapable. A radiation probe will never sit in complete silence; it will emit a never-ending chatter of *click click clicks* no matter where it is. You can't have 'radiation-free' anything. Background radiation is about as ubiquitous and as harmless as it gets.

High altitude and uranium-rich bedrock in Colorado give inhabitants a radiation dose of 6.1 millisieverts, twice the US average. Millions of Coloradans spend their entire lives living (and breathing) in the enhanced radiation, and yet their cancer incidence rate is 11 per cent lower than the national average.[11]

Exceptionally thorium-rich sands in Kerala, south India, make it one of the most radioactive places on Earth. Background radiation in some areas is 70 millisieverts, 30 times higher than the global average.

A survey of 70,000 Keralites published in 2005 found no increased risk in cancer related to high background radiation.[12]

But nowhere is more naturally radioactive than Ramsar in northern Iran. In some parts of the city, background radiation is 260 millisieverts, more than 100 times the global average. If I had a sample at Sellafield that exposed me to that much radiation, I'd be obliged to put special precautions in place. And yet, a synthesis of the scientific literature in 2019 found no evidence that the elevated background-levels of radiation increases cancer rates.[13] If they do cause cancer, the effects are so small that they're yet to be proven.

Environmental traces

On top of natural background radiation, there are a few sources of radiation we, humans, have put into the environment that contribute to our annual dose.

Atom bomb tests in the mid-twentieth century blasted smatterings of highly enriched uranium, plutonium, and fission fragments high into the atmosphere. Wind carried the nuclear fallout – so-called because it literally 'falls out' of the sky – around the planet. At its peak in the mid-1960s, the fallout added 0.11 millisieverts to the global annual dose, or an extra 17 days' worth of natural background radiation. But after more than half a century of laundering by the weather and radioactive decay, it's faded to a trifling 0.005 millisieverts, or 18 hours' worth.[14]

Traces left by Chernobyl persist in the environment, too. Fires lofted motes of spent fuel high into the atmosphere, which settled across much of the northern hemisphere. Average dose in the northern half of the planet following the accident was 0.04 millisieverts annually, or an extra 6 days' worth. Today it's a mere 0.002 millisieverts, just 7 hours' worth.[15]

Nuclear power also leaves its traces. Mining and milling uranium ore, fabricating fuel rods, running nuclear power stations, and dealing with the waste inevitably release tiny amounts of radioactivity into the environment. That radioactivity adds about 0.0002 millisieverts annually to the global average, or 45 minutes' worth. Those living within a mile or so of a nuclear power station can expect slightly more, about 3 days' worth.[16]

Incidentally, coal contains traces of radioactive isotopes that are released into the environment when we burn it. Per kilowatt-hour of electricity, coal power gives rise to 60–140 per cent more radiation exposure than nuclear power. Shutting down nuclear power stations and replacing them with coal-fired power stations in a bid to reduce the public's dose would be spectacular self-sabotage.[17]

So how does it all add up? Suppose you lived in the northern hemisphere down the road from a nuclear power station (which, as it happens, I do). Your annual dose from atom bomb fallout, Chernobyl debris, and nuclear power would be no more than the harmless natural background you'd receive by spending a day or two in Colorado or Cornwall. It's utterly inconsequential, and no ground on which to reject nuclear power.

※ ※ ※

Shortly after Becquerel discovered radioactivity, it became clear that substantial exposure to alpha and beta particles is harmful. In a 1927 lecture, Rutherford would recall a dinner party in 1903 with Pierre and Marie Curie. After what Rutherford described as a 'lively evening', they retired to the garden, where Pierre Curie produced a vial of radium solution coated in zinc sulphide. The zinc sulphide fluoresced white from the alpha and beta radiation. 'The luminosity was brilliant in the darkness and it was a splendid finale to an unforgettable day.' But Rutherford's fond recollections had a foreboding underdone: 'At the time we could not help observing that the hands of Professor [Pierre] Curie were in a very inflamed and painful states due to exposure to radium rays.'[18] In his inaugural Nobel Lecture in 1905, Pierre Curie attested to the ill effects of alpha and beta radiation:

> If one leaves a wooden or cardboard box containing a small glass ampulla with several centigrams of a radium salt in one's pocket for a few hours, one will feel absolutely nothing. But 15 days afterwards

a redness will appear on the [skin], and then a sore which will be very difficult to heal ... Radium must be transported in a thick box of lead.[19]

I wince when I read these accounts. Gram for gram, radium is 1.4 million times more radioactive than natural uranium.* Radiation's ability to spark cell mutations was unknown in the early 1900s, but by the time Marie Curie died of leukaemia in 1934 aged 66 – a cancer almost certainly caused by her massive and prolonged exposure to radiation , although she had exceeded the life expectancy in France at the time by several years[20]—— the link was crystallising. But a vast cavern separates harmless background radiation from having a glowing vial of radium in your pocket. This begs the question you've been itching to ask: *how much radiation is too much?*

Alas, it's difficult to answer because estimating risk is notoriously complicated. Consider the risk of cancer brought about by smoking cigarettes. You might reasonably ask how much your risk of lung cancer increases if you smoke 4, 14, or 40 a day. But it's impossible to say with absolute certainty. Some 90-year-olds smoked a pack every day of their lives; non-smokers occasionally die of lung cancer aged 20. There's obviously a massively increased risk of lung cancer brought about by smoking cigarettes, but it's difficult to quantify with *certainty*. We can only estimate, and therefore must satisfy ourselves with *statistics*.

The relationship between radiation and cancer has been subject to intense research for more than a century. In 1946, American geneticist Hermann Joseph Muller won the Nobel Prize in Physiology or Medicine for 'the discovery of the production of mutations by means of X-ray irradiation'. His subjects? Fruit flies. He blasted them with varying fluxes of X-rays and measured their incidence of genetic mutation. In his inaugural Nobel Lecture, Muller outlined the state of knowledge at the time:

---

* Radium is an alpha-emitter, and so wasn't to cause of Pierre Curie's sores; the walls of the glass ampulla, and the cotton of his shirt pocket, would have blocked the alpha particles. His burns were caused by radium's beta-emitting radon progeny.

...the frequency of the gene mutations is directly and simply proportional to the dose of [radiation exposure] . . .[21]

In other words: the more radiation you're exposed to, the more likely you are to get cancer. Muller continued:

[There is] no escape from the conclusion that there is no threshold dose . . .

In other words: every single time you're hit by an X-ray, gamma ray, beta particle, or alpha particle, you're rolling the dice of death; an increased risk of cancer follows each increment of dosage, no matter how small, and there is no amount of radiation that is completely safe.

Muller's summary – a *linear* relationship between radiation and cancer with *no threshold* at the lower end – yields the *linear no-threshold model*. This has become the orthodoxy upon which essentially all radiation protection laws and regulations are based. Muller's principle dictates how I handle samples in the lab every day; it decides how many X-rays a physician will take of your broken bones; it influences how nuclear power stations are run and how spent fuel is stored.

We can plot the linear no-threshold model to life on a graph, with *increase in cancer risk* on the vertical axis, and *radiation dose* on the horizontal axis.

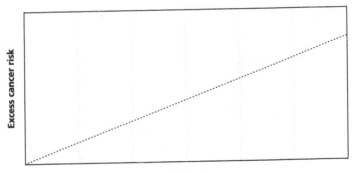

Note that the trendline is straight. That's the *linear* part of the model. Also note that it goes directly through the origin, where there's an increased cancer risk all the way down to zero. That's the *no-threshold* part.

Now, don't be fooled by the sure-looking graph. The linear no-threshold model is just that – a model. It's a simplified reflection of messy reality that is, at best, an approximation. I'm reminded of the aphorism, *all models are wrong, but some are useful*. That the trendline underpins radiation protection policy does not mean it is scientifically robust. Policy decisions have no weight in scientific arguments.

So, exactly how useful is the linear no-threshold trendline? This happens to be one of the most controversial topics in modern science. Delving into the literature on the subject feels like entering a war zone. In 2015, John Boice Jr – former president of the US National Council on Radiation Protection and Measurements – quipped, 'LNT [linear no-threshold] is not TNT, but differences in opinions sometimes appear explosive!'[22]

But there's good reason for the controversy. To understand, let's look beyond the model at some data.

Testing the model

At least one thing is for sure: practically everybody agrees on the top-right corner of the linear no-threshold trendline: the largest radiation doses are bad news for human health. And by 'largest', I mean upwards of a few thousand millisieverts in one go. (Recall, background radiation is a few millisieverts annually.) Under these circumstances, cells are damaged in such devastating numbers that the body's repair mechanisms are overwhelmed. The effects manifest as *acute radiation sickness*. Symptoms vary from person to person, but they generally encompass a searing fever, gut-wrenching sickness, and pervasive malaise.[23] I've heard it been likened to three dimensional sunburn. If the dose is large enough, death ensues before cancer even has a chance to develop.

One such incident happened in the summer of 1945. Harry Daghlian – a physicist working at the Los Alamos research facility in New Mexico – was experimenting with a fist-sized sphere of metallic plutonium. He was stacking bricks of tungsten carbide around the sphere to measure their neutron-reflecting properties; by bouncing neutrons back into the plutonium, went the idea, criticality could be achieved with small amounts of fissile material. Daghlian crafted his precarious contraption by hand. As he lowered the final brick towards the plutonic sphere, his neutron counters indicated it would send the sphere super-critical. He withdrew the neutron-reflecting brick . . . but it slipped from his grasp and clattered onto the plutonium. *Super-criticality.* Stray neutrons, beta particles, and gamma rays burst forth from the sphere. A blue flash illuminated the lab as the air itself was stripped of its electrons. Daghlian immediately knocked the tungsten carbide bricks away and severed the fission chain reaction. But it was too late. Ten million billion atoms had fissioned within a split-second. The air offered no protection from the intense flash of energy, and he received about 5,100 millisieverts, or about 2,100 years' worth of background radiation all in one go. Within 30 minutes, his fingers were tingling; within 15 days, they'd succumbed to gangrene and necrosis. Radiation sickness took hold. A worsening fever pushed him into a coma from which he never awoke, and he died on the 24th day.[24]

Death within a fortnight is inevitable for poor souls exposed to more than 10,000 millisieverts. Between 2,500 and 5,000 millisieverts, the odds of dying within a few months are 50-50. Below a few thousand millisieverts, the chance of a rapid death plummets, but an elevated risk of cancer lingers after the acute radiation sickness subsides.[25]

Exposures to such profuse bombardments of radiation only happen in the most extreme circumstances. By themselves, these rare incidents cannot help us test the validity of the linear no-threshold trendline because the data are too scattered. We need to furnish them with larger datasets. One such bank of data exists.

In August 1945, the USA exploded atom bombs over two Japanese cities: a uranium-235 bomb over Hiroshima, and a plutonium-239 bomb over Nagasaki. At least 180,000 people died within 4 months of the blasts. The true death toll is unknowable. Also unknowable are the precise causes of death; traumas, severe burns, and acute radiation sickness afflicted hundreds of thousands of people simultaneously, making it impossible to parse the details from the calamity. Some 284,000 survivors were exposed to varying levels of radiation from the explosions.[26]

After World War II, a spirit of cooperation flourished between Japan and the US. Born from that cooperation was a scientific endeavour called the Life Span Study. The alliance – the first of its kind – monitors the long-term health outcomes of atom bomb survivors. By 1950, it encompassed 120,000 individuals, 94,000 of whom were in Hiroshima or Nagasaki when the bombs exploded. The other 27,000 weren't in either city at the time of the bombings, and so act as an unexposed control group. In 1958, the systematic monitoring of their health outcomes began. Their lives have since been placed under a microscope.[27]

Crucially – and what makes the Life Span Study so special – scientists meticulously reconstructed each person's radiation dose based on exactly where they were at the time of the detonations. Reconstructions even went into such details as whether they were indoors or outdoors, and whether they were facing away from or towards the explosion. Doses ranged from more than 2,000 millisieverts all the way down to background levels. Almost half of the individuals were exposed to less than 5 millisieverts, which is less than the natural annual background radiation in the Czech Republic or Finland.[28]

Scientists then tracked individuals over the intervening years and decades to see how many got cancer. They also collected data on lifestyle – such as smoking, drinking habits, and obesity. The Life Span Study is the largest and longest-running of its kind, widely considered to be the gold standard for discerning the effects of radiation on long-term health outcomes.[29]

The most immediate effect of the radiation was an excess in leukaemia deaths amongst survivors. Other types of cancer – such as breast, thyroid, and intestinal – took slightly longer to manifest.[30]

The probability that an individual succumbed to this fate depends on their radiation exposure. By combining cancer deaths and dose estimates, we can furnish our cancer *versus* radiation graph with real data and test the validity of the linear no-threshold trendline. This is what our graph, based on 53 years of Life Span Study data, looks like:

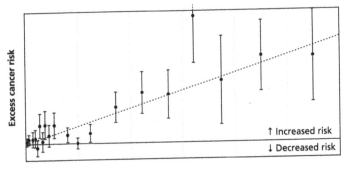

Increase in cancer risk versus radiation dose, furnished with real data collected between 1950 and 2003 in the Life Span Study. I omitted units for clarity because it's the trend that matters.[31]

There exists a consensus in the scientific community on the general trend: at large doses, the more radiation you're exposed to, the more likely you are to get cancer. The rule of thumb is that for every thousand millisieverts, your chance of dying from cancer increases by about 5.5 percentage points. To put that in context, consider the EU, where cancer causes 22 per cent of all the deaths (the remaining 78 per cent die of other causes); that means if you picked 1,000 Europeans at random and followed them through their entire lifetimes, you'd find 220 would die from cancer. If you'd exposed a similar group to 1,000 millisieverts, there would be an *additional* 55 cancer deaths.[32]

But beyond general trends, things unravel. Sloping down the trendline towards low doses – which is where almost all exposures lie – there's so much noise in the data that it's impossible to pin down the true trajectory. This is where the linear no-threshold model crumbles, giving way to controversy.

Don't panic

The trendline below a hundred or so millisieverts is fuzzy. Supplementing the Life Span Study with data from other sources – like those exposed at Chernobyl – adds little in the way of clarity.

Some studies claim the linear no-threshold model is accurate at low doses, and that the trendline really does follow a straight trajectory to zero. But a few studies are more pessimistic, claiming that small radiation exposures are disproportionately worse than high ones. You can see hints of that in the datapoints that lie above the trendline. And a growing number of studies claim the contrary, presenting evidence that small doses are actually good for you because they reduce your risk of cancer; this hypothesis is called *radiation hormesis*, and you can see hints of it in the datapoints that fall off the bottom of the graph. Scientists argue passionately with one another as they attempt to tease truth from statistics. But we just don't know for sure. Even if tiny amounts of radiation are harmful, the effects are too small to see, even when the health outcomes of more than 100,000 people are analysed forensically. The data are utterly inconclusive.[33]

This calls the two eponymous features of the model into doubt: it may not be *linear* all the way down, and there may be a *threshold* – a hundred millisieverts, or thereabouts – below which there are no impacts on health.

The model also has another fundamental flaw: it treats all radiation as cumulative, making no distinction between sudden exposures, and those that happen over long stretches of time. For instance, we know that 500 millisieverts spread over an entire lifetime – as somebody in Finland would receive naturally – is less harmful than 500 millisieverts all in one go.[34]

Thus, the health effects of radiation below a hundred(ish) millisieverts are unknown. And that in itself is telling. It means there's no obvious harm for the doses encountered by the vast majority of people. If harm does exist, then it's slight enough to be elusive after decades of raging scientific debate. The radiation experienced by practically everybody is of no measurable consequence.

※ ※ ※

Every year, we perform more than 4 billion medical X-rays in hospitals around the globe. Spread across the entire population, they add 0.6 millisieverts to the average global dose, which is roughly the same as spending a month in Colorado. Medical examinations are humanity's largest source of radiation exposure after radon.[35]

I fractured my arm falling from a rope swing when I was six. I did the same to my wrist when I fell off a skateboard aged 14. On both occasions, radiographers X-rayed my broken bones and exposed me to about 0.001 millisieverts. That's an extra 0.05 per cent on top of annual dose from background radiation, equivalent to a fleeting 3.5 hours' worth.[36]

Dental X-rays are slightly more potent, exposing patients to about 0.03 millisieverts (4 days' worth); chest X-rays more so, at 0.1 millisieverts (a fortnight's worth); and mammograms even more, at 0.36 millisieverts (about 8 weeks' worth). A CT scan of your whole spine – about 9 millisieverts – exposes a patient to about 4 years' worth of average background radiation, or the same as what somebody living in Finland receives naturally every 18 months.[37]

What about flying aboard an aeroplane? Aboard a 9-hour transatlantic flight, you'd be exposed to about 0.09 millisieverts, or about 3 dental X-rays' worth. I once flew from Glasgow to Budapest to present on plutonium chemistry at a conference; I was exposed to more radiation on the flight than I was during all the hours I spent tinkering in the lab.[38]

What about flying aboard a spacecraft? Orbiting the Earth 250 miles above us, the International Space Station sails through an almost-perfect vacuum. Up there, there's no atmosphere to attenuate the downpour of cosmic rays. Astronauts are exposed to 0.4 millisieverts every day; that's the equivalent of an entire year's worth of earthly background radiation in less than a week.[39] Mortality

Research & Consulting, Inc. recently compiled mortality data for all NASA astronauts since 1959 and all Soviet or Russian cosmonauts since 1961. Their research encompasses more than 10,000 person-years of follow-up time. Some astronauts did die from cancer, as you'd expect, but there was no statistically convincing excess. It was as though they'd never been to space, despite their high doses.[40]

Since I started working at Europe's biggest nuclear site, I've done chemistry with uranium, neptunium, plutonium, and americium; I've stood in the vicinity of fission products, weighed crumbs of enriched uranium powder, and stood within diving distance of spent fuel storage ponds. My annual dose averages out at 0.3 millisieverts. That's the same as I'd receive every 18 hours aboard the International Space Station. Put another way, it's the same as spending about a fortnight in Cornwall, or taking a couple of transatlantic flights. At the rate I'm going, I'll tot up a spinal CT scan's worth of radiation before I retire.

It's often people who work with radiation every day – nuclear scientists, radiographers, power station workers, *et cetera* – who are the least troubled by radiophobia. Through silver screens, visions of blooming mushroom clouds, and sometimes deliberate fearmongering, popular culture has warped societies' perception of radiation. The reality is far less dramatic: radiation, for practically everybody on Earth in practically every imaginable situation, is nothing to worry about. And, reassuringly for our net zero efforts, it's no ground on which to reject a massive expansion in nuclear power.

Chapter 9. We Need To Talk About Chernobyl

When you invent the ship, you also invent the shipwreck; when you invent the plane you also invent the plane crash; and when you invent electricity, you invent electrocution . . . Every technology carries its own negativity, which is invented at the same time as technical progress.

— Paul Virilio, philosopher and cultural theorist, 1999

It was early in the morning on 28 April 1986. Cliff Robinson — a scientist working at Sweden's Forsmark Nuclear Power Plant, two hours north of Stockholm — had just eaten his breakfast in the staff lounge. After brushing his teeth, he was ready to start his shift. On his way back to the power station's locker room, he walked through a radiation detector. It burst into an alarming wail. The soles of his shoes were, apparently, radioactive, even though he hadn't stepped foot in the controlled area.

Robinson tried again, and the siren rang once more. He made a third attempt, but this time, he was met with silence. He shrugged it off as a false alarm. Sometimes the detectors were over-eager. Thinking nothing more of it, he went about his shift as normal. But when he returned later in the day, he found a long queue of perplexed workers. The detector wasn't letting anybody through because the alarm kept ringing out. Everybody, apparently, was contaminated. Something was wrong.

Puzzled, Robinson borrowed a shoe from one of the people in the queue. He took it to the lab and placed it on a gamma-ray spectrometer. 'Then' Robinson recalled, 'I saw a sight that I will never forget': sharp peaks — the unmistakable signals of gamma rays emanating from radioactive atoms.

Fearing radioactive dust had somehow escaped the reactor, Forsmark was put on high alert. Engineers and scientists forensically combed the facility looking for a breach in containment. Robinson was amongst them: 'Nothing indicated any malfunction or problem at Forsmark; it was just that the surroundings were very heavily contaminated.' They couldn't find anything wrong with their reactor. Everything was intact. They would soon learn their attempts were in vain. The radioactive dust wasn't coming from Forsmark. It wasn't even coming from inside Sweden. It was coming from a reactor some 770 miles to the southeast that had caught fire two days earlier. The burning reactor was one of four at the Vladimir Llyich Lenin Nuclear Power Plant, better known simply as *Chernobyl*.[1]

The world got its first inkling of the nuclear disaster unfolding behind the Iron Curtain from the sole of a shoe. The discovery of fallout in Sweden – traced back to the Soviet Union by the southeasterly winds on which it sailed – forced the Soviet authorities to admit what had happened. They were arrogant in thinking they could have kept the secret for long; atoms care little for national borders.

Nuclear's reputation was incinerated by Chernobyl, and radiophobia continues to smoulder. It effectively ended Europe's expansion of nuclear power; it built more reactors in the 5 years leading up to Chernobyl than it has in the decades since. In 2023, the Public Attitudes Toward Clean Energy Index – the world's largest multinational opinion poll on nuclear power – surveyed more than 20,000 people spanning 20 nations. Four out of every five respondents were either *fairly concerned* (31 per cent) or *very concerned* (49 per cent) about nuclear's safety record.[2]

Over-moderated and out of control

The reactors at Chernobyl were RBMK reactors,* built by Soviet engineers in the 1970s and 1980s. And they're total oddballs (the reactors, not the engineers).

---

* The acronym RBMK comes from the Russian *реáктор большóй мóщности канáльный* – '**r**eaktor **b**olshoy **m**oshchnosty **k**analny' – meaning *high-power channel reactor*.

RBMK reactors were powered by slightly enriched uranium (about 1.8 per cent uranium-235) and cooled by light water. Remember: whilst light water moderates neutrons a little – it facilitates links in fission chain reactions – it also simultaneously dampens the nuclear furnace by absorbing some of them. The combination of light water and uranium that's only slightly enriched isn't potent enough to elevate k above 1 and send the reactor critical.[3]

Pretty much every other nuclear reactor in the world gets around this problem in one of two ways: either by using *light* water as a moderator but burning uranium *enriched* to something like 4 or 5 per cent uranium-235; or by using *heavy* water as a powerful moderator and burning natural, *non*-enriched uranium as a fuel.

Soviet engineers, however, built a lattice of graphite blocks around the RBMK core. Graphite is such a good neutron moderator that it can send slightly enriched uranium critical (or even natural uranium, à la Fermi), even when it's dampened by light water. The combination of a graphite moderator and light-water coolant is unique to RBMK reactors. No other type of power reactor in the world uses it.

And that combination is an issue. Whilst most reactors have *negative temperature coefficients* – a self-regulating negative feedback loop stops the reactor overheating – RBMK reactors can have *positive* temperature coefficients. This means, under certain conditions, the fission chain reaction ramps *up* as the core's temperature increases, and runs the risk of going super-critical and out of control.*[4].

In the small hours of 26 April 1986, reactor operators were performing a planned safety test in Chernobyl's Unit 4. But they went rogue, recklessly pushing the RBMK reactor far beyond what it was ever designed to do by flouting strict operating procedures. This precipitated an enormous power surge in the core. Fuel rods ruptured as super-heated water flashed to steam. The overpressure exploded the reactor with enough force to lift the 1,000-tonne lid.

---

* My job here is to examine the consequences of Chernobyl rather than the details of its causes. For an excellent account of what happened and why, I recommend the 'Chernobyl Accident 1986' page on the World Nuclear Association's website.

Boom. Incandescent lumps of spent fuel and irradiated graphite burst forth and set the surrounding buildings ablaze. The flames lofted radioactive dust high into the skies of the Soviet Union. Radioactive particles rained upon modern-day Ukraine, Belarus, and Russia. Two days later, Cliff Robinson's shoes would be contaminated in Sweden. In time, the plume would touch the entirety of the northern hemisphere.[5] Chernobyl, ultimately, was the result of a quirky reactor design, rule-breaking operators, and corrupt Soviet leadership.

And so came to pass the worst disaster in the history of nuclear power.

Aftermath

The first fatalities were a pair of power station workers. The explosion killed them immediately.[6]

Next were the firemen and power station workers. Many of them came into direct contact with chunks of spent fuel fresh from the reactor's core. Acute radiation sickness afflicted 134 of them because the shrapnel, crammed with fission fragments, was screaming with radiation. Their doses ranged from 800 to 16,000 millisieverts, equivalent to between 330 and 6.5 thousand years' worth of background radiation. Over the months that followed, 106 recovered and 28 died.[7]

But what about the long-term effects? Latent cancers are difficult to count precisely, but agencies systematically monitored the health outcomes of the 106 survivors. Like the atom bomb survivors of the Life Span Study, they had their lives placed under a microscope. By 2007 – more than two decades after Chernobyl exploded – 19 of them had died: 14 died from illnesses unrelated to radiation, such as heart attacks (6), tuberculosis (2), and trauma (1). The remaining 5 died of cancer. Exactly how many of the cancer deaths were caused by Chernobyl is unknowable. The United Nations Scientific Committee on the Effects of Atomic Radiation – a multinational group of scientists that synthesises the best available science on the health effects of radiation – concede that as time passes, 'the assignment of radiation as the cause of death has become less clear'. This leaves the confirmed death toll of people directly involved with the accident somewhere in the low-to-mid-30s.[8]

That's about the same number of people who die at work in the United Kingdom every 3 months. Admittedly, they aren't normally killed by an exploding reactor; they mostly die by falling from height or being run over. Every death from Chernobyl was an avoidable tragedy, and so by making such comparisons, we run the risk of sounding callous. But it's important we put numbers into context. Thirty or so immediate deaths at Chernobyl was 30 or so too many, but it's a surprisingly small toll given the mythical status of the night in question.[9]

In the aftermath of the explosion, some 600,000 people – the so-called *Chernobyl liquidators* – were drafted from across the Soviet Union to clean up the surrounding area. The grim task took them 3.5 years. Despite the liquidators spending months in the immediate vicinity, long-term health studies have failed to find a link between their radiation doses and cancer fatalities.[10]

But what about latent cancers amongst the wider population? There was a concerted effort to monitor the long-term health outcomes of those exposed to fallout. But mercifully, radiation doses were comparable to that of the natural background for the vast majority. Even in contaminated areas of Ukraine, Belarus, and Russia, the average additional dose over the 20 years following the accident was around 9 millisieverts, or an extra 3.75 years-worth of natural background radiation (the same as a CT scan). A 2019 study in the *Journal of Global Oncology* found that cancer incident rates in the regions of Ukraine close to Chernobyl were no higher than the national average.[11]

The only cancers with an unambiguous link to Chernobyl are thyroid cancers, caused by a type of atom with an unfortunate combination of radioactive and biochemical properties: iodine-131.[12]

Iodine-131 is a common fission fragment. Its 8-day half-life makes it – gram for gram – 180 billion times more radioactive than natural uranium. It's also readily absorbed by the human body, especially by the thyroid. When people ingest iodine-131, their thyroid absorbs it with the same eagerness as it absorbs non-radioactive, regular iodine. Authorities normally hand out iodine tablets to areas affected by nuclear fallout; by flooding the human body with a surplus of non-radioactive iodine, the thyroid's hunger is sated and it ignores the foreign iodine-131. But the Soviet response to Chernobyl was slow. Iodine-131 worked its way into the region's milk supply after being eaten by the cows that grazed on contaminated pastureland. Then it made its way into humans. As it silently fired out beta particles and gamma rays, it exposed its host to a dose of radiation.[13]

There's no evidence of an increased incidence of thyroid cancer amongst those who were adults at the time of the accident. But there *was* an increase amongst those who were children or adolescents. By 2005, there had been 15 confirmed deaths.[14]

In the 24 years between 1991 and 2015, there were just shy of 20,000 cases of thyroid cancer across Ukraine, Belarus, and Russia amongst those who were under 18 at the time of Chernobyl. In 2018, the UN Scientific Committee on the Effects of Atomic Radiation linked about a quarter of those cases to iodine-131. (This estimate carries a large uncertainty, and the true portion could be as low as 7 per cent or as high as 50 per cent.) And so we might expect something like 5,000 cases of Chernobyl thyroid cancer in the long run, but it could be anywhere between 1,400 and 10,000. We can combine this estimate with thyroid cancer survival rates to calculate the number of fatalities: the prognosis for childhood thyroid cancer is excellent; survival rates lie somewhere in the region of 97 to 99 per cent. Therefore, we might expect something like 100 (or so) deaths from Chernobyl-induced thyroid cancer. Even in a pessimistic scenario – where 50 per cent of registered thyroid cancers were down to iodine-131 – we'd expect a few hundred Chernobyl-induced thyroid cancer fatalities at most.[15]

Thus, there are about 50 confirmed deaths from Chernobyl: 30 in the immediate aftermath, some latent cancers amongst the first responders, plus the 15 thyroid cancer fatalities. The *true* death toll – which we cannot quantify perfectly – is higher because of thyroid cancers that are yet to present themselves, but it likely falls in the region of a few hundred. Whilst imprecise, these calculations give us a feel for the magnitude of Chernobyl's death toll.

The data stand in stark contrast to the widespread belief that

Chernobyl sent a wave of cancers careening through Europe. In 2011, the Union of Concerned Scientists – a staunch anti-nuclear campaign group – claimed that Chernobyl will eventually cause 12,000 to 57,000 fatal cancers. But their estimates are based on the dubious linear no-threshold model, which, as we saw in the previous chapter, lacks convincing evidence. Such grim outlooks are reinforced in popular culture, too. Against a backdrop of sombre choral music, HBO claimed in the closing credits of its 2019 *Chernobyl* TV miniseries that 'there was a dramatic spike in cancer rates across Ukraine and Belarus'. Thankfully, the data show these gloomy diagnoses to be false.[16]

Today, the Chernobyl exclusion zone is a tourist attraction. (Or it *was* a tourist attraction until it was placed off-limits by Russia's 2022 invasion of Ukraine.) Guides offer private tours, and you can even rent your own dosimeter. Spending an hour in the vicinity of the decrepit power station will bathe you in about 0.005 millisieverts, which is about six times less than a dental X-ray.[17]

Mind over atoms

Beyond the small number of fatal radiation-induced cancers, a far more insidious affliction crept over the people affected by Chernobyl: mental health issues. The chaotic relocation of some 350,000 people from the exclusion zone left a scar on their collective psyche. Liquidators and 'exposed persons' – regular citizens and responders alike – were stigmatised. Massive upheaval and pervasive radiophobia amplified cases of chronic stress, depression, anxiety, and post-traumatic stress disorder. In 2020, the World Health Organization concluded that 'the biggest health impact of Chernobyl has been on mental health'. Mothers with young children evacuated from the exclusion zone were particularly afflicted: they suffered increased rates of depression and post-traumatic stress disorder because of fears about the effects of Chernobyl on their children's health.[18]

Expectant mothers were also impacted. Radiophobia motivated thousands of agonised women to seek abortions out of fear of radioactive contamination and foetal birth defects. In the months following Chernobyl, abortion rates spiked as far away from the disaster as Denmark, Italy, and Greece. But those fears were misguided. Radiation doses were so low amongst the affected populations that Chernobyl fallout had no measurable effect on fertility, the frequency of stillbirth, or the incidence of birth defects. 'Not surprisingly, none [of the babies], at birth at least, has any detectable abnormalities,' a physician from the University of California at Los Angeles told journalists a year after the accident. 'We weren't expecting any. That's an example that no news is good news.' Thousands of would-be parents mourned needlessly.[19]

Suicidal thoughts, depression, and alcoholism continue to disproportionately haunt liquidators compared to the wider population. Researchers tracked the health outcomes of 4,812 Estonian liquidators over the 35 years after the accident and found no radiation-induced excess cancer risk. But they *did* find an excess in cancers caused by smoking and excessive drinking. They also found excessive suicide mortality. The researchers concluded that 'the psychosocial impact [of Chernobyl] was greater than any direct carcinogenic effect of low-dose radiation'.[20]

All these psychological effects were intensified by the turmoil that followed the disintegration of the Soviet Union 5 years after the accident. Characterising affected populations as 'victims' rather than 'survivors' exacerbates the pervasive feeling of helplessness and despair.[21] Radiophobia, in the end, killed more people than radiation.

※ ※ ※

At 14:46 on 11 March 2011, Japan's main island of Honshū lurched eastwards by almost 2.5 metres, and 400 kilometres of its coastline dropped by 2 feet. It was the most powerful recorded earthquake in Japan's history. The seismic shift displaced colossal amounts of ocean water and sent a tsunami hurtling towards the archipelago's Pacific-facing coastlines.[22]

Japan lies close to the intersection of two tectonic plates. It's no stranger to earthquakes and so had built automatic shutdown mechanisms into its 54-strong fleet of nuclear reactors. At 14:47 – one minute after the earthquake struck – control rods autonomously plunged into the cores of Units 1, 2, and 3 at the Fukushima Daiichi Nuclear Power Plant. (Units 4, 5, and 6 were already switched off and down for maintenance.) Neutron poison severed the fission chain reactions.

The spent fuel in the reactors brimmed with the radioactive heat of decaying fission fragments. No matter, though: pumps continued to circulate cold water through the core to prevent the fuel from overheating.

At 14:48 – one minute after the reactors shut down – Fukushima lost power. The earthquake had triggered a landslide that knocked over one of the pylons carrying electricity to the reactor site. Just as the cooling pumps whirred down, a small army of back-up diesel generators sprang to life, electrifying the power stations and returning cold water to the core. Everything went exactly as planned.

At 15:27 – 41 minutes after the earthquake struck – the first tsunami made landfall. The seawall held it back. But 10 minutes later, the second tsunami, bigger than the first, hit. A 43-foot swell of ocean water, crowned with white foam and spray, poured over the seawall and inundated the power station. And then things stopped going to plan. The back-up diesel generators were overwhelmed. A bank of back-up batteries kept the cooling pumps turning for a few hours, but they soon drained. Fukushima lost power, the cooling pumps spluttered to a standstill, and the cores were cut off from their supply of cold water. The devastation caused by the tsunami prevented emergency workers from restarting the pumps. Thus began the worst nuclear incident since Chernobyl.[23]

The temperature inside the reactor cores began to rise. Within hours, the uranium fuel – white-hot with fission fragments – melted. The searing decay heat coaxed a chemical reaction between the zirconium (Zr) in the fuel cladding and the water (H_2O) in the core. Oxygen atoms shuffled themselves into kinship with zirconium, yielding zirconium oxide (ZrO_2) and hydrogen gas (H_2), the latter of which is explosive.[24]

Exactly 24 hours after the tsunami breached the seawall, the free hydrogen gas combined violently with atmospheric oxygen in Unit 1. *Boom.* The reactor building exploded, sending a cloud of radioactive dust into the air. Two days later, the same thing happened in Unit 3. *Boom.* Meanwhile, hydrogen gas from Unit 3 flowed through ductwork and accumulated inside Unit 4. The day after Unit 3 exploded, Unit 4 followed suit, blasting out a third radioactive cloud. *Boom.*[25]

A hole in the side of Unit 2 – probably caused by the flying shrapnel from Unit 1 – acted as a pressure-relief valve through which hydrogen could escape, thus preventing its explosion. Unfortunately, the hole allowed even more radioactive dust to escape into the environment.[26]

Placing the back-up diesel generators below the high-water mark was a painful mistake.

Overreaction

Over 20,000 people perished in the earthquake and subsequent tsunami. It ranks amongst the worst natural disasters of the twenty-first century. But nobody died from the exploding nuclear reactors at Fukushima. Nor did any first responders suffer acute radiation sickness in the aftermath.[27]

What about latent deaths amongst the first responders and power station workers? In the years since the nuclear meltdown, a single death has been linked to radiation – a man who died from lung cancer in 2018. Radiation doses amongst first responders were so low that future radiation-induced cancers, should they transpire, are likely to be indiscernible from background cancer rates. That brings Fukushima's confirmed direct death toll to 1.[28]

And what about latent deaths amongst members of the public? There aren't any. In 2022, the UN Scientific Committee on the Effects of Atomic Radiation synthesised 11 years of Fukushima research and found no evidence that radiation had caused an increase in any type of cancer. In fact, they found no evidence that radiation had caused *any* adverse health effects at all, even amongst those who lived close to the power station.[29]

It's no surprise there was so little collateral damage. Amongst the most exposed members of the public – infants who lived within

20 to 30 kilometres of the power station – radiation doses were less than 8 millisieverts over the year following the accident. The worst-affected people living in the Fukushima prefecture will be exposed to no more than about 17 millisieverts over their entire lifetimes. To put that into perspective, Coloradans are exposed to 7 millisieverts of natural background radiation every single year – or some 550 millisieverts throughout their lifetime – from natural background radiation.[30]

Thus, the final death toll from Fukushima radiation is unlikely to exceed 1. (We can't rule out more in the future; it's possible there will be an accident whilst the clean-up operation is ongoing.) But as was the case with Chernobyl a generation before, there was a far bigger, indirect killer: radiophobia.

International guidelines limit the public's radiation dose from nuclear power stations to 1 millisievert per year. These guidelines are set by the International Commission on Radiological Protection. But by the International Atomic Energy Agency's own admission, the guidelines are 'prudent' and based on a dubious hypothesis that has 'not yet been established' – the linear no-threshold model. If we extended these guidelines to include natural background radiation, we'd have to evacuate at least 20 US states and half of the countries in Europe.[31]

In adherence with the stringent guidelines, the Japanese authorities ordered the mass evacuation of civilians who lived near the power station. Some 160,000 people were displaced. Scores were crammed into temporary shelters, gymnasiums, and school halls. It's crucial to keep in mind that Japan was handling the evacuations whilst coping with the aftermath of the earthquake and tsunami; resources were stretched. Either way, the physical stress and emotional turmoil of the evacuation killed 2,313 people. Nine in ten of those deaths were amongst the elderly.[32]

Nearly all the evacuations were unnecessary, however, because most of the evacuation zone had radiation levels no higher than the natural background. The evacuations were largely an overreaction motivated by the paranoid 1-millisievert limit. 'With hindsight,' Philip Thomas – Professor of Risk Management at the University of Bristol – said, 'we can say the evacuation was a mistake. We would have recommended that nobody be evacuated.'[33]

Heroic efforts over the years following the accident saw cleanup workers scrub down buildings, wash the streets, and scoop up 9 million cubic metres of contaminated earth. Most of the region surrounding Fukushima is today less radioactive than Colorado, North Dakota, and Cornwall, even across vast swathes of the 20-kilometre exclusion zone. And yet, by 2020, fears of radiation had kept about 38,000 evacuees from returning home.

There's a strong argument that rapid and extensive evacuations were entirely justified as a precautionary measure in the most heavily contaminated areas. It's also an open question as to exactly how many radiation-induced deaths they prevented. But most of the evacuees should probably have been kept away for weeks, not years.[34]

Thus, the final confirmed death toll of Fukushima is 2,314: 1 from radiation, plus 2,313 from the evacuation. Shizuyo Sutou – a professor at Shujitsu University – said in 2016, 'The most threatening and debilitating public health issue is the adverse effect on mental health caused by undue fear of radiation.' I'm reminded of the aphorism: *history doesn't repeat itself, but it often rhymes*.[35]

Fukushima water

In the immediate aftermath of Fukushima, power station workers flooded the reactor cores with water to quench their heat. Rainwater and groundwater subsequently infiltrated the broken reactor buildings, adding to the spate. The water inherited some of the spent fuel's radioactivity when it became ladened with enriched uranium, a medley of fission fragments, and a smattering of transuranics. The radioactive water accumulated in thousands of giant storage tanks, and by 2023, there were 1.3 billion litres of it. That's enough to fill more than 500 Olympic-sized pools.[36]

Japanese authorities laboriously cleansed the contaminated water by drawing it through a sequence of filters and chemical scrubbers. They siphoned away most of the radioactive isotopes. But there was one unstable isotope that proved impossible to remove: hydrogen-3, a beta-emitter better known as *tritium*. The technology to chemically purify tritium from water in such profuse volumes doesn't exist, and with a half-life of just over 12 years, it would take more than a century for the tritium's radioactivity to diminish a thousandfold. That's far too long for it to sit in temporary storage, especially as contaminated water continues to accumulate. The Japanese authorities were left with one option: to discharge the radioactive water offshore and let the world's vastest ocean dilute it into insignificance.[37]

In August 2023, Japan pressed ahead and released the first tanks of contaminated water into the Pacific. It was more a trickle than a surge – it will take until at least the early 2070s to discharge all of them. Nonetheless, Japan's neighbours cried out in condemnation. The Chinese government publicly accused Japan of 'passing an open wound onto the future generations of humanity' and decried its abandonment of 'moral responsibilities'. Protestors in South Korea and Hong Kong marched through the streets holding toy fishes tattooed with canary-yellow ☢ symbols. Greenpeace alleged the Japanese government was disregarding scientific evidence and violating human rights.[38] But whilst the radiophobia is real, the danger posed by the water is non-existent.

A pint of Fukushima water contains 2 trillionths of a gram of tritium, which is 6.5 times below the safe drinking water limit set by the WHO. Once it's piped offshore, the contaminated water is immediately diluted by enormous volumes of ocean water. The radioactivity thins almost into nothingness. Even in an extreme scenario – where a person inhales copious amounts of Fukushima sea spray, frequently eats platters of Fukushima seafood, and enjoys swimming a few miles offshore of the Fukushima power station – the annual radiation dose would be hundreds of times less than a single dental X-ray.[39]

Of all the radioactive isotopes produced as a byproduct of nuclear power, tritium is one of the most benign. Its beta radiation carries such feeble energy that it can't even penetrate skin, and so it doesn't inflict much in the way of radiation dose. It also has a short residence time in the human body. After 5 weeks, 90 per cent of any ingested tritium is excreted naturally as it's replaced by normal hydrogen in food and drink. A colleague once told me that three power station workers at Sellafield were accidentally exposed to large amounts of tritium back in the 1960s. All three of them were, apparently, marched down to the nearest pub and told to drink as much lager shandy as possible to speed up the hydrogen-excretion process. The tritium was flushed from their bodies before any harm could be done. I'm pretty sure my colleague was joking – he's always pulling my leg – but it illustrates the principle.[40]

If Japan really is violating human rights by releasing the water, then those violations pale in comparison to those perpetrated by Mother Nature. A perpetual pitter-patter of tritium falls naturally from the sky, created when cosmic rays collide with atmospheric nitrogen atoms and smash them to pieces. The tritium latches on to oxygen atoms to form mildly radioactive water (sound familiar?) before raining into Earth's water cycle. Most of it becomes part of the oceans; some of it becomes part of *you*. The tritium released into the Pacific Ocean from Fukushima annually is the same as what's synthesised naturally in the sky every few hours.*[41]

The Chinese government's accusation that the Fukushima water represents 'a disaster to the local people and the whole world' and is an 'extremely selfish and irresponsible act' is shamelessly hypocritical, anyhow. Japan releases about 0.06 grams of tritium into the ocean from Fukushima every year. China discharges 50 times that amount from its own nuclear power stations.[42]

It's perfectly possible to pipe radioactive water offshore without playing politics with radiophobia. France sends about 30 grams of tritium – 500 times the Fukushima annual discharge limit – into the English Channel every year from La Hague. It has no adverse impacts on the local ecosystem or marine environments, and the annual dose to humans is less than 0.00001 millisieverts, equivalent to a few minutes' worth of natural background radiation.[43]

Most of the United Kingdom's radioactive water comes from its biggest nuclear site, Sellafield. Every year, it flushes about half a gram of tritium into the Irish Sea, about 10 times the Fukushima

---

* Tritium doesn't accumulate naturally on Earth over time because it radioactively decays as quickly as it's cosmogenically synthesised. The decay and synthesis balance each other out; there's only ever about 5.5 kilograms of natural tritium on the entire planet at any one time.

annual discharge limit, along with a pinch of plutonium and a thimbleful of fission fragments. (Sellafield discharged about 10 times that amount until its two La Hague-style fuel recycling facilities shut down in in 2018 and 2022.) Even amongst the most exposed members of the public – like those who eat lots of local seafood and spend lots of time on nearby beaches – the radiation is a hundred-fold less than average background radiation. During the summer months, alongside thousands of others, I brave the cold British waters down the coast from Sellafield for a swim. And I haven't grown gills, yet.[44]

Cold mathematics

Here's a blunt calculation, but bear with me: for a given electricity source, take the total number of people it's killed and divide that number by the total amount of electricity it's generated. This lets us compare the death rate – expressed in deaths per terawatt-hour – of different electricity sources, like for like. It's morbid, but it's necessary if we're to form cogent perceptions of safety. In 2022, Hannah Ritchie from *Our World in Data* synthesised the numbers and performed the calculation:

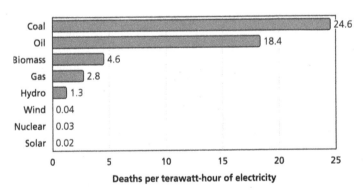

Deaths per terawatt-hour of different electricity sources. Wind, nuclear, and solar are so safe that they're invisible on this scale.[45]

There's some wiggle room in these numbers. The calculated death rates of coal, oil, and gas are vast underestimates; the fossil fuel calculations are based on European data, where power stations and the air they dirty tend to be further away from cities than in other parts of the world. Deaths from wind and solar come with a fair degree of uncertainty because they aren't documented systematically. Nuclear's death toll doesn't include the impossible-to-quantify deaths from mental disorders after Chernobyl and Fukushima. But the numbers represent best estimates and reflect the overall picture: nuclear is about as safe as wind and solar, and it's tens or hundreds of times safer than fossil fuels and biomass. Nuclear's death rate would have to be underestimated by 9,200 per cent to make it as deadly as gas.

The nuclear calculation includes immediate and long-term deaths from Chernobyl and Fukushima. It would have included the deaths from the partial meltdown at the Three Mile Island – after which the USA didn't build another reactor for 35 years – but there weren't any. Nuclear's safety record is blotted by a small number of rare, high-visibility events. The uncommon fatalities are diluted by the enormous quantities of power it produces. But nuclear *feels* far more deadly because of the radiophobia that pervades society.[46]

Hydro's death rate is similarly dominated by a rare, large-scale accident: the catastrophic failure of the Banqiao Dam in China in 1975, which collapsed after Supertyphoon Nina dumped more than a year's worth of rain in less than a day. A wall of water – 9 metres high in places, and 12 kilometres wide – inundated an area half the size of Wales as it surged downstream at 25 miles per hour. The precise death toll is unknown because the Chinese Communist Party attempted to cover up the calamity. But best estimates place it at 26,000. Some 145,000 more died in the squalor and famine that followed. Banqiao aside, hydro is about as safe as wind, nuclear, and solar.[47]

You might be wondering how wind turbines and solar panels kill people: they're deaths by freak accidents, such as when ice falls from turbine blades and strikes people stood beneath, or when engineers get electrocuted and fall from rooftops whilst they install panels. When compared like for like – terawatt-hour for terawatt-hour – wind and solar are as benign as atom splitting.[48]

Let's put those death rates into context by imagining a city of 1 million average OECD citizens, using their 47 megawatt-hours of energy annually. If we powered the city using nothing but coal, we'd expect about 22 deaths per week. If we replaced coal with gas, we'd expect two or three deaths per week. With hydro, we'd expect just one death per week. But if we powered our city with wind, nuclear, or solar, we'd have to wait more than six months for a single fatality. I know where I'd rather live.

Bad air

Each year, some 8.8 million lives are cut short by air pollution. That's a higher annual death toll than all natural disasters combined (tens of thousands), HIV/AIDS (hundreds of thousands), and COVID-19 (an average of 1.6 million per year during the first 4.5 years of the pandemic). Dirty air ranks as the world's fourth-biggest killer after tobacco, high blood pressure, and malnutrition. It's one of the greatest global health catastrophes of our time.[49]

By far the biggest polluters of the air we breathe are the fuels we burn. In 2019, a landmark study estimated that some 3.6 million people die every year as a direct consequence of outdoor air polluted by fossil fuels. Those deaths stem mainly from power stations, industry, and cars.[50]

Nuclear power, by contrast, does not cause air pollution. The clouds of white 'smoke' you might see billowing from the fat chimneys of a nuclear power station aren't smoke at all – they're steam. It's just water. By replacing fossil fuels with uranium, the air becomes cleaner. Nuclear power stations saved an estimated 1.8 million lives between 1971 and 2009 because of the fossil fuels and the downstream air pollution they displaced.[51]

Following Chernobyl, engineers modified the USSR's 16-strong fleet of RBMK reactors to fix their safety issues. Those reactors have collectively totted up almost 500 years' worth of running time since. Seven of them are (at the time of writing) still running.[52] In the decades since Chernobyl, those RBMK reactors – once considered to be some of the most dangerous nuclear reactors in the world – have killed nobody. There isn't a single fossil fuel-fired power station in the world that can make the same claim.

In addition to deaths from outdoor air pollution, indoor air pollution kills a further 3.8 million people annually. That's because 3 billion people – more than a third of the global population – have no choice but to cook using dirty fuels such as wood, charcoal, and kerosene. The smoky fires kindled by these fuels release noxious pollutants that cause all manner of fatal diseases. People living in low- and middle-income households are the most afflicted.[53]

Air pollution from fossil fuels kills as many people every 6 hours as nuclear power has killed ever. Its death rate is equivalent to a Chernobyl every 30 minutes. Dirty air doesn't look as nightmarish as Chernobyl liquidators, but it's orders of magnitude more deadly. Fission, all deaths considered, is far safer than setting things on fire.

It's important to acknowledge that deaths from air pollution aren't the poisoned fruits of malevolence. Nor are they the tragic outcome of ignorance. They're borne from necessity. Air pollution deaths are the consequence of our species' utter dependence on energy. But whilst they're unintentional, many are avoidable, because we have the means to replace fossil fuels with cleaner energy sources. There are no trade-offs in this respect: the source of energy with the lightest environmental burden – nuclear – is also one of the safest.

Unintended consequences

Spooked by Fukushima, Japan embarked on a rapid nuclear winddown in the years following the earthquake. Nuclear power stations generated a quarter of Japan's electricity in 2010. In 2014, they generated nothing. The shortfall was filled almost entirely by fossil fuels, including coal. The carbon-intensity of its electricity rose by a third as a direct consequence.[54]

The increase in coal power dirtied Japan's air, too. Researchers at Columbia University estimate that substituting nuclear power for coal caused 10,000 to 27,000 additional air pollution deaths in Japan between 2011 and 2017. Meanwhile, almost 6,000 miles away, Fukushima motivated Germany's *atomausstieg*; and the deadly air pollution arising from its heightened dependence on coal killed an estimated 5,600 people between 2011 and 2019.[55]

As a volcanic islands chain, Japan is not endowed with coal seams or gas fields, and so it relies heavily on imports. That makes fossil fuels expensive. The financial burden of replacing nuclear power with fossil fuels fell on bill-payers, making electricity unaffordable for many in the wake of Fukushima. Citizens were forced switched off their electric heating units to save money, and in the 3 years following the accident, an estimated 4,500 people died from the cold.[56] All those deaths dwarf the fatalities from the accident at Fukushima itself.

Whilst Germany's nuclear phase-out is supposedly permanent, Japan's rev-down turned out to be a hiatus. It restarted a pair of 900-megawatt pressurised water reactors at the Sendai Nuclear Power Plant on Kyūshū island in 2015. Eight years later, it had restarted another 10 reactors that collectively fulfilled 8 per cent of its electricity needs. It plans to restart more in the coming years.[57]

A decade after Fukushima, Hiroshi Kajiyama – Japan's Minister for Economy, Trade, and Industry – said 'nuclear power will be indispensable' if the country is to reach its 2050 net zero target. A cold snap of low winds had brought Japan close to blackouts in 2021. 'Solar wasn't generating. Wind wasn't generating. I'm trying to persuade everybody that in the end we need nuclear power,' Kajiyama said. Japan aims to generate 20 to 22 per cent of its electricity from nuclear by 2030, which would place it just shy of where it was before Fukushima. It was also amongst the signatories of the Declaration to Triple Nuclear Energy in 2023.[58]

⚛ ⚛ ⚛

Nuclear's safety record isn't a reason to reject it. On the contrary: in return for a death rate comparable to wind and solar, we get energy that's abundant, reliable, and clean. It's a trade-off with maximum return and minimum impact on human health. Rashly replacing nuclear power stations with fossil fuels costs lives. The hysterical opposition to nuclear power distracts us from the preventable dangers that kill millions annually and ruins our best shot at renouncing fossil fuels before 2050.

It's not the case that nuclear power is perfectly safe. Nothing is. But splitting atoms in power stations is one of the safest things our species does.

Chapter 10. Golden Geese

Atomic-energized seeds are jolted with gamma rays of nuclear energy!
Little things or great things can happen. The possibilities are fantastic!

– A newspaper ad, *Chicago Tribune*, 1962

Since its discovery, we've used radiation to probe the inner workings of material reality. Those early experimentalists – Becquerel, Rutherford, the Curies, Meitner, Hahn, Strassmann, Frisch, amongst thousands of others – used radiation to uncover the structure of atoms. And by treating it with due care and respect – and a dash of creative flare – scientists and engineers today use radiation to solve problems in novel ways.

We use radiation to sterilise medical equipment, for instance. More than 160 nuclear facilities around the world blast scalpels, syringes, and other instruments with gamma rays to kill germs. Since gamma rays shine through almost everything, they even sterilise equipment that's already been packaged. More than 40 per cent of all single-use medical devices in the world are cleaned in this way.[1]

We also blast food with radiation to keep it fresh for longer. Gamma rays from radioactive isotopes such as cobalt-60 and caesium-137 turn germs like *Salmonella* and *E. coli* into toast. They also delay ripening and sprouting, giving more time for food to find its way to our dinner plates before it spoils. Astronauts eat meat that's been sterilised by radiation to avoid food poisoning whilst 250 miles above the nearest hospital.[2] Preserving food with radiation is the nuclear equivalent of pasteurising milk.

We manufacture 460 *million tonnes* of plastic every year. Shamefully, we recycle less than 10 per cent of it. The rest is either incinerated, sent to landfill, or dumped. Plastic pollution is an environmental catastrophe. One of the problems is the sheer difficulty of recycling it; plastic is designed to be chemically tough, which makes it awkward to break down and remake into something else. But chemical bonds are no match for radiation. By blasting plastics with radiation, we can break their molecules into simpler pieces that are easier to recycle. We already use this technology to recycle old car tyres into roofing tiles and asphalt, and Teflon into powders for paints and industrial lubricants.[3]

Radiation acts as a local fireguard, too. You've likely got a few hundred nanograms of radioactive americium-241 inside your smoke alarm. (Incidentally, that americium-241 is probably the heaviest type of atom in your house, and it's likely your smoke alarm is the most radioactive object you own.) As it transforms into its neptunium-237 progeny, americium-241 spits out alpha radiation. That radiation electrifies the air in its vicinity and permits an electric current to flow. When smoke obstructs the alpha radiation – remember, alpha particles are easily blocked – the circuit breaks, sounding the alarm. One of the health physicists I work with told me a former colleague of his used to test his radiation probe with the office smoke alarm: he'd pop off its cover, hold the probe against the americium-241, and wait to be met with those *click click clicks*.

Americium-241 also emits weak gamma rays as it decays, which shine silently into your house: the extra radiation exposure over your entire lifetime is equivalent to a fleeting few hours' worth of natural background.[4] It's a trade-off worth making, given the protection it provides against a fire.

A determined person could increase their exposure by disassembling their smoke alarm and eating the kernel of americium-241. Even then, their extra radiation exposure would equal just 2 or 3 months' worth of background radiation. A paper in the journal *Health Physics* from 1977 describes a bizarre case where lady who worked in a smoke alarm factory accidentally swallowed not one but *two* specks of americium-241 simultaneously. (Exactly how she came to swallow the transuranic morsels isn't detailed.) Researchers tracked the specks as they passed through her body, as a metal detectorist scans the ground for hidden treasure. She 'voided' – as the paper euphemistically puts it – the first speck after 16 days, and the second speck followed 8 days later. A detailed assessment of her radiation dose during her ordeal concluded the health impact of the radiation was 'not significant', which I suppose is comforting on the off chance your smoke alarm falls from your ceiling and into your breakfast.[5]

I would advise against snacking on americium-241, though. In fact, I'd advise against eating anything from the nuclear world ... unless, that is, it was grown in an atomic garden.

Atomic gardening

The evolution of life is driven by genetic mutations. Some mutations are unfavourable because they curtail an organism's fitness; the natural environment selects against them, thereby eliminating them and their potential offspring from the gene pool. Now and then, though, a mutation bestows some advantage over its fellow specimens. The favourable trait propagates onwards through its hereditary line, thus tweaking – or *evolving* – the character of the species.

Homo sapiens is the only form of life that consciously directs the course of evolution to its own advantage. By permitting organisms to reproduce when they possess traits we deem favourable – normally edibility or usefulness, but sometimes cuteness – we mould species into forms we see fit. This process is called *domestication* or *selective breeding*, but I guess you could call it *unnatural selection*. We've been doing it for tens of millennia. It's how we crafted golden retrievers from wolves, pigs from wild boars, and chickens from junglefowl.

It's also how we domesticated the plants that underpin modern agriculture. For 12,000 years, our ancestors sowed seeds, nurtured shoots, and tilled the soil. Only the seeds from the best plants were sown in the next harvest. They selectively cultivated the crops that grew easily, delivered the biggest bounties, and possessed the most nutrition. Everything we eat is now genetically modified.

Although our selection was *unnatural*, the genetic mutations that underpinned our endeavours depended wholly on nature's hand. We relied on the spontaneous scrambling of genetic code caused by hiccups in DNA replication. It involved lots of waiting. Progress was millennia in the making. But in the early twentieth century, we acquired a tool by which we could mimic nature by artificially scrambling genetic codes and thus accelerate the rate of mutation: *radiation*.

The Fantastic Four became superheroes when a freak storm of cosmic rays caught them off-guard; Spider-Man acquired his superpowers when a radioactive spider bit him; and the Incredible Hulk attained his super-human strength when a gamma-bomb blasted him. Whilst super-humans are confined to the comic book universe, super-*plants* – bred by exposing them to intense radiation beams – are real. Our combination of radiation science and a green thumb is, delightfully, called *atomic gardening*.

Gamma gardens started springing up everywhere following Eisenhower's 'Atoms for Peace' speech. In their atomic allotments, scientists arrange thousands of crops, cuttings and seeds in concentric rings about a central plinth. Atop the plinth sits an intensely radioactive lump, normally of synthetic cobalt-60. It's quite an easy isotope to concoct inside nuclear facilities: simply bombard normal cobalt – which is all cobalt-59 – with neutrons and it occasionally absorbs one. Whilst normal cobalt isn't radioactive, cobalt-60 is, gram for gram, about 1.6 billion times more radioactive than natural uranium, with a short half-life of 5.25 years.[6] It decays by firing out a high-energy beta particle and, crucially, an intense gamma ray. The gamma rays shine from the cobalt-60, showering the plantstuff with tens or hundreds of thousands of millisieverts.

After a thorough dousing, scientists remotely lower the cobalt-60 from the plinth into a lead-lined bunker. Then, they reap their harvest. Most plants perish, especially those close to the garden's centre. But further away, radiation exposure ebbs and genetic mutations abound. Thousands of plants acquire undesirable traits (or none at all), but *some* mutate into botanical delights. There's no way of predetermining what traits an irradiated plant will gain. You have to grow it and see what happens.

The International Atomic Energy Agency and United Nations curate a list of atomic plants on the Mutant Variety Database. It sounds like something from a 1950s monster flick, but, at the time of writing, it contains more than 3,400 entries (or should I say *entrées*?).

Now, don't let radiophobia fritter away your appetite: atomic gardening does *not* make plants radioactive. In fact, it doesn't leave any discernible mark of human intervention. It simply does what nature does – create offspring with new traits – but faster.[7]

Amongst the nuclear treats listed in the Mutant Variety Database is a strain of rice bred by Chinese scientists; they found their crops became resistant to certain types of fungi, grew faster, and yielded a bumper harvest after they'd sprayed them with 200,000 millisieverts of gamma rays. After irradiating pollen from tomato plants with 23,000 millisieverts, Japanese scientists found their plants were more resistant to disease, less likely to suffer root rot, and needed less water to grow. And scientists in Mauritius found their oyster mushrooms grew faster and tasted better after a dousing from a whopping 400,000 millisieverts. A note on the database describes them as 'well appreciated when cooked' and 'firm and tasty'.[8]

Many a floral delight has sprouted from gamma gardens, too. Using gamma rays, scientists in the USA created a variety of fungi-resistant rose with two-tone pink petals. No fewer than eight fresh varieties of tulip have come from atomic gardens in the Netherlands, with characteristics including new colours, patterned variegations, and better bulbs. And scientists in India changed the colour of their gladioli from purple to pink.[9]

And on a more whimsical note, a gamma breed of barley acquired higher yields and enhanced malting properties when British scientists zapped it with 240,000 millisieverts of gamma rays in the mid-1960s, making it ripe for a spot of brewing.

Atoms for peanuts

By the late 1950s, atomic gardens were blooming in research facilities across Asia, India, North America, and Europe. Meanwhile, the idea of atomic gardening had taken root in the public, and amateur nuclear enthusiasts were eager to share in the thrill. Whilst they couldn't build DIY nuclear reactors in their sheds, they *could* tend to homemade gamma gardens. Such was the pastime of Tennessean dentist Clarence

Speas. After obtaining a licence to own and use radioactive isotopes, he ordered 9 thousandths of a gram of cobalt-60 to his house.* The crumbs arrived through the post, stowed safely inside a third of a tonne of lead. In his backyard, Speas bathed millions and millions of seeds in the hope of breeding new and wonderful varieties of plants. The atomic begonias, chrysanthemums, and roses he displayed at flower shows caused a minor media storm. With an entrepreneurial streak, Speas sold his 'atomic energized seeds' to would-be atomic gardeners across North America, running sensational newspaper ads with the promise of fantastic results. He even offered a cash prize to gardeners who grew the strangest new plants.[10]

> *Dr. Speas, ATOMIC-ENERGIZED SEEDS & PLANTS!*
> *The possibilities are absolutely fantastic.*
> *Be one of the first to actually grow unpredictable 'Atomic' Flowers &*
> *Vegetables. You just might originate a totally new variety—WORTH*
> *MONEY to you! Join the world-wide search for NEW plant species.*
> *No guarantees—only NATURE knows what will happen, but one thing's sure:*
> *You'll enjoy untold gardening fun!*

If I were to start my own line of atomic seeds, I'd give them element-themed names. I'd turn delicate geraniums into 'uraniums', pretty tulips into 'neptulipiums', and tasty yams into 'sweet plutatoes'.

By autumn 1960, Speas had shipped more than 3.5 million seeds across the ocean to the UK. One of his biggest customers was Muriel Howorth.

Nuclear energy sparked Howorth's curiosity when she was 62 years old. Unfamiliar with the science of atom splitting, Howorth borrowed a book from the library written by Frederick Soddy: *The Interpretation of Radium and the Structure of the Atom*. Soddy had won the Nobel Prize in Chemistry in 1921 for discovering that elements have different forms: *isotopes*. On the book's 183rd page, Howorth read the words:

---

* I couldn't believe it when I learned this. If I handled that amount of cobalt-60 in my fume hood at work, I'd exceed the allowable radioactivity limit.

A race which transmute matter would have little need to earn its bread by the sweat of its brow . . . such a race could transform a desert continent, thaw the frozen poles, and make the whole world one smiling Garden of Eden.[11]

(Ironically, nuclear power would *stop* us thawing Earth's frozen poles by halting climate change.) 'By the time I had reached page 183 I was a firm believer in the power of the atom for good,' Howorth would later recall.[12]

Convinced that humanity's fortune would be elevated by nuclear technology, Howorth founded the Atomic Energy Association of Great Britain in 1948 to bring nuclear science to the layman. Her ambition didn't stop there; she wanted to bring nuclear science to the lay*woman*, too. Striving to (in her own words) 'lead woman out of the kitchen and into the atomic age', tireless Howorth also founded the Ladies' Atomic Energy Club. 'Not to know all about atomic energy and the wonderful things it can do is like living in the Dark Ages,' she declared.[13]

Glowing with atomic enthusiasm, Howorth also penned and produced a pantomime called *Isotopia: An Exposition on Atomic Structure*. It included characters such as Isotope, Neutron, and Atom Man. (Members of the Ladies' Atomic Energy Club played all the parts.) A *Time* magazine review of the performance in 1950 describes how cast members in 'flowing evening gowns gyrated gracefully about a stage in earnest imitation of atomic forces at work'.

Howorth aspired to perform *Isotopia* at the Royal Albert Hall in London. 'We would have room there', she explained to a journalist through a smile, 'for all the 92 transmutations of the atom.'[14] Alas, *Isotopia* basked in the limelight for one night only.

In 1959, Howorth hosted a dinner party at which she served peanuts. But they weren't just any old peanuts. They were *North Carolina 4th generation X-rayed* peanuts, made giant by a radiation-induced mutation. They dwarfed the regular peanuts she served alongside. But to her dismay, the salted marvels of atomic gardening were lost on her guests: '[They didn't] seem to appreciate that the nut was . . . the outcome of an immense achievement'. Howorth vented her frustration with a spot of gardening. She '[popped] an irradiated peanut in the sandy loam to see how this mutant grew'.

Four days later, Howorth's peanut had sprouted. Her plant was soon 2 feet tall. Delighted (and probably a bit smug), she called the newspapers, and within days she was caught in the swell of a media frenzy. Newspaper and television interviews ensued, and journalists queued down her driveway as they craned for a closer look at the marvellous plant.

Beverley Nichols – a beloved celebrity gardener of the day – described it as 'the most sensational plant in Britain' that 'had all the romance of something from outer space'. The humble peanut plant embodied the spirit of atomic gardening: 'It holds a glittering promise in its green leaves, the promise of victory over famine'.

Riding the wave of fame, Howorth inaugurated the *Atomic Gardening Society*. Members received six gamma-blaster peanuts when they signed up.[15]

Sadly, the Atomic Gardening Society is no more. As the post-war decades ensued, enthusiasm for the peaceful applications of nuclear science was eclipsed by fears of the atomic arms race. Public interest wilted. Although atomic gardening now takes place in research institutions rather than backyards, it continues to fortify the global food supply and reverse the socioeconomic fortunes of poor, rural communities.

Today, in Peru, atomic breeds of barley like *UNA-La Molina 95* – which slides easily out of its indigestible husk – and *Centenario II* – which is high-yield and hail-resistant – mean Andean farmers now grow enough food to meet their own needs *and* sell the surplus to market. They've even built small factories that process the barley into flour and pearls. In recognition of her work, leading gamma-breeder Professor Gomes Pando won the Peruvian Prize of Good Governmental Practices in 2006.[16]

Over the past five decades, scientists at the Bangladesh Institute of Nuclear Agriculture have bred no fewer than 13 new varieties of atomic rice. They helped Bangladeshi farmers triple their rice production between 1987 and 2017. Suruj Ali, a farmer from Gerapacha village in northern Bangladesh, grows a variety of atomic rice called

Binadhan-7. It's fast-growing, possesses enhanced disease resistance, and – according to the Mutant Variety Database – is tasty. 'I have more rice for my family, and I now earn almost double with the rice and mustard seed I grow, compared to before,' said Ali. His rice crops thrive with less intervention from pesticides, too, which is good for his pocket and the local wildlife. 'I've used that extra money to build two new extensions for my house. I hope I can earn enough to send my kids abroad someday.'[17]

Atomic rice also elevates the fortunes of Vietnamese farmers. According to scientists from Vietnam's Institute of Agricultural Genetics, it's produced 'remarkable economic and social impacts, contributing to poverty alleviation in some provinces'.[18] With atomic gardening we grow more food using less resources. It's good for people *and* the planet.

Debugged

In the late 1930s, insect scientists Edward Knipling and Raymond Bushland – nicknamed 'Knip' and 'Bush' – were working for the United States Department of Agriculture. They were devising new ways of combating the legions of insects that infect animals and decimate livestock. Their target was the New World screwworm. It was ravaging herds of cattle everywhere from the southern states all the way down into South America.

The New World screwworm is a metallic-blue fly that lays its eggs in the wounds of warm-blooded animals. It takes its name from the maggots that hatch from those eggs, so-called because they burrow, screw-like, into the skin of their host. The maggots feast on the living flesh with their sharp mouth-hooks. Their point of entry need not be larger than a pinprick, but the wound widens as it festers, exposing flesh on which even more flies can lay their eggs. The maggots munch deeper and deeper in search of fresh food. The host is eventually overwhelmed. It's an excruciating and often fatal infestation. Livestock, wildlife, humans, and pets are all susceptible. A single female can lay 3,000 eggs in less than a month. The screwworm has earned the second term of its Linnaean binomial, *Cochliomyia hominivorax*, 'eater of man'.[19]

Needless to say, New World screwworms inflict misery, not to mention massive economic damage. The universe would be a better place without their insectile pestilence. By 1960, screwworm infestations in livestock were costing the USA $80 million – more than $800 million in today's money – every year.[20]

In devising schemes to lessen their numbers, Knipling made an important observation: female screwworms are monogamous, mating just once in their 3-week lifespan. Male screwworms, on the other hand, go about with unrestrained promiscuity. This gave the scientists an idea. If they could somehow render a huge fraction of the male screwworm population infertile, they'd trick the females into wasting their one and only mating opportunity on a seemingly healthy, but secretly sterile, partner. They'd sever their reproductive cycle. The screwworms would unwittingly eradicate themselves.[21]

Interfering in the sex lives of New World screwworms sounds eccentric. Indeed, Knipling and Bushland's idea was met with ridicule: 'You just can't castrate enough flies,' one of their colleagues quipped. And I guess they had a point – you can't castrate billions of flies all at once. Knipling and Bushland needed a way of mass-producing infertile screwworms quickly and cheaply.

The solution came from Hermann Joseph Muller, the father of the linear no-threshold model, who had published work describing how radiation exposure had rendered some of his fruit flies infertile. Inspired, Knipling and Bushland thought they could use radiation to sterilise male screwworms in droves. They ran their idea by Muller, who wrote back: 'I know nothing of screwworms, but your theory is sound.'[22]

Knipling and Bushland bred captive screwworms in Petri dishes on a diet of lean beef. And then they blasted the juveniles with 50,000 millisieverts of X-rays using medical equipment in a nearby army hospital. The pupae matured to flyhood apparently unharmed. But the radiation had made them infertile. Irradiated males could not fertilise eggs; irradiated females could not produce eggs that hatched. And crucially, exposure to X-rays did not quench the passions of the licentious males, which went about their business as if nothing had ever happened. It was time to test the idea in the real world.[23]

Their testbed was Sanibel Island, a thin strip of land just off the Floridian coast. There, screwworms lived off a diet of tormented cats, opossums, and rabbits. In 1951, scientists began airdropping sterile screwworms onto the island at a rate of 100 per week for every square mile of the island. The sterile males mingled with the island's indigenous females, which were – unbeknownst to them – destined to lay eggs that would never hatch. After 8 weeks, authorities declared screwworms eradicated. It worked ... until un-irradiated screwworms flew over from the continent.[24] It was time to try again, far from the mainland.

Their island of choice was Curaçao, some 50 miles off the coast of Venezuela, where screwworms plagued the islanders' goats. The experiment began in spring 1954. Scientists flew sterile screwworms from a mass-production facility in Orlando, Florida, and airdropped them at a rate of 200 per week for every square mile of the island. By August, they'd increased the drop rate fourfold. One of the scientists later recalled how they 'experienced great suspense each evening as we tabulated the egg mass data'. Four months later, screwworm was declared eradicated.[25]

If it were possible to rid islands of the pestilence, why not entire states?

Government authorities turned to Florida, where screwworm infestations cost farmers about $200 million (in today's money) every year. They intensified their production of sterile screwworms in factory-sized radiation facilities. Output surged from 200,000 flies per week to more than 60 million. It was an all-out screwworm blitz. Ten aeroplanes began airdropping the infertile screwworms in May 1958, and the final infestation was reported 9 months later. It worked. The Sunshine State was free, and it only cost $100 million.[26]

If it were possible to rid entire states of the pestilence, why not entire countries?

Authorities dropped irradiated screwworms from the skies in their billions. It was screwworm Armageddon. In 1966, the US

Department of Agriculture declared screwworm eradicated across the entire US mainland. Authorities pushed the screwworms deeper into Central America, rolling the drop zone southward like a slow-motion carpet bomb. Eradication was declared in Mexico in 1991, Belize and Guatemala in 1994, El Salvador and Honduras in 1996, Nicaragua and Costa Rica in 1999, and, finally, Panama in 2006.[27]

Today, Panama rears more than a billion sterile screwworms annually inside its Sterile Fly Producing Plant. They're raised on a tailored diet of eggs, milk, and animal blood before being blasted with radiation. Six flights a week airdrop them into the wilderness at Panama's southern border.[28] The wild population cannot pass north through the drop zone into Central America, as the females mate with sterile males and leave no offspring to continue north. The invisible barrier prevents screwworms crossing back into Central and North America. The institution upholding the endeavour – the Panama-United States Commission for the Eradication and Prevention of Screwworm – uses the outline of a fly set inside the silhouette of an atom as its emblem.

In spring 1988, strange maggots were discovered feasting on the festering wounds of livestock in Libya, North Africa. Nothing of their kind had been seen before in the region. Samples were sent to the Natural History Museum in London, where taxonomists had a grim realisation: it was the dreaded screwworm. It had made its way as a transatlantic stowaway. The infestation spread rapidly over a vast area, threatening the entire Mediterranean and the whole of Africa, and by 1990, more than 12,000 cases were confirmed. But that's where this horror story ends. Nuclear science – mobilised through the Food and Agriculture Organization of the United Nations – stopped a nightmare from unfolding. More than 1.3 billion irradiated screwworms were flown across the Atlantic and airdropped over North Africa, and by the end of 1991, the scourge had been eradicated. Unchecked, the screwworms would have devastated livestock in some of the world's poorest regions, inflicted untold suffering, and decimated Africa's unique wildlife. The entire thing cost a measly $64 million and saved an estimated $1 billion in economic damage; that's a saving of more than $15 for every dollar spent.[29] The environmental savings – which include the continued existence of some of Earth's most beloved animals – are priceless.

Screwworms were just the start. Radiation has also been used to eradicate the tsetse fly – which spreads the fatal Trypanosome parasite – from parts of Africa; it's been used to suppress a flurry of pestilent moth species; and it's been deployed in battle against various species of fruit flies – which ravage fruit and vegetables – on every continent bar Antarctica.[30] All these programmes avoid the collateral damage to the local ecosystem wrought by spraying insecticides.

Efforts are underway right now to use radiation sterilisation against one of humanity's oldest enemies: the mosquito. It's responsible for transmitting almost 1 in 5 of all infectious diseases globally. Amongst those diseases are dengue fever, of which half of all people are now at risk, and malaria, which kills more than 1,300 children per day, mostly in the world's poorest regions. Under the guidance of several major institutions – including the International Atomic Energy Agency – scientists are figuring out how to send mosquitos the same way as screwworms and fruit flies.[31]

Rhisotopes

As well as eliminating the small insects that plague us, we can use radiation to save the giant animals that enthral us.

In March 2018, the world's last male northern white rhino – a handsome chap called Sudan – passed away at his home at the Ol Pejeta Conservancy in Kenya. He left behind the last two members of his kind: Najin and her daughter, Fatu. When Najin dies, Fatu will be the last of her kin. And when Fatu dies, the northern white rhino will vanish from the face of the planet forever.[32]

A clue to the northern white rhino's demise lies in Najin and Fatu's horns. Or rather, the stumps where Najin and Fatu's horns once proudly stood. Conservationists sawed them off to deter the rhino's greatest peril: poachers. Further south, poachers have slashed Africa's southern white rhino population by a third since 2012 in merciless pursuit of their horns. In 2017, poachers broke into a Parisian zoo in the dead of night and shot a four-year-old white rhino called Vince thrice in the head before hacking off his horn with a chainsaw. The closely related black rhino remains critically endangered because of poaching, too. The Javan rhino is also critically endangered; poachers slaughtered Vietnam's last one in 2010.[33]

The decimation of the rhino is fuelled by a lust for their horns, mainly in Asia. Through holes in airport security, smugglers slip them into the East, where they're sold on black markets as ornaments of wealth and ingredients in traditional medicines. Practitioners claim powdered rhino horn can do everything from ease a hangover to cure cancer. The whole thing is quackery. It doesn't deter the poachers, though, because – gram for gram – it sells for more than gold and diamond. Poachers, smugglers, and buyers mostly get away with their crimes.[34]

Najin and Fatu will spend the rest of their lives under the protection of armed guards, 24 hours a day. It's a symbol of the brutal arms race being fought across Africa against the poachers, who relentlessly pursue the gentle herbivores with night vision cameras, helicopters, and tranquillisers.[35] They generally don't grant their victims the relief of death before they start up their chainsaws.

It's an appalling situation. But every so often, the brightest ideas emerge in the darkest circumstances. This is where Doctor Lorinda Hern enters our story. Poachers murdered a rhino cow – heavily pregnant at the time – and her 2-year-old calf on Hern's family game reserve in 2010, making off with the horns. From her anguish came a clever concept: if she could somehow make rhino horns worthless, the financial incentives for poaching would vanish. She started investigating ways of devaluing rhino horns by spiking them with wildlife-friendly poisons, dangerous for human consumption. She founded two charities, Rhino's Last Stand and Rhino Rescue Project, with the aim of doing just that.

In 2019, Hern had another stroke of inspiration: instead of devaluing rhino horns by making them merely poisonous, let's render them utterly worthless by making them radioactive. Who wants to own radioactive contraband? She put her idea to her friend Professor James Larkin, the director of the Radiation and Health Physics Unit at the

University of Witwatersrand in South Africa. Larkin is an expert in radiation control and nuclear security, and a big animal lover.[36]

'My initial reaction when Lorinda asked me what I thought was *no!* I'd spent my entire career protecting people from radiation. But then I started to think . . . *what if?*' Larkin told me.

Thus, the Rhisotopes project was born. Larkin has since recruited a team under a single goal: to save the rhino using nuclear science. They plan on pushing metal pins — made radioactive by infusing them with 15 to 20 billionths of a gram of cobalt-60 — deep into the horns of living rhinos. The tiny pins will be screaming with beta particles and gamma rays, acting as a silent nuclear deterrent.

Cobalt-60 would make rhino horns dangerous to ingest. (Fifteen-billionths of a gram of cobalt-60 contains as much radioactivity as 34,000 bananas.) Their extra dose of gamma rays means they couldn't even sit atop a mantlepiece as a monstrous ornament. 'We want to use people's fear of radiation against them,' Larkin told me with a wry smile, 'which is funny, because I've spent my career teaching people not to be afraid of radiation.'[37]

There's another advantage to making rhino horns radioactive. Severed rhino horns are normally invisibly stowed inside suitcases, but the radiation would make them beam like beacons. Larkin explained, 'There are somewhere between 10,000 and 11,000 radiation detectors installed in airports around the world.' Horns would be far less likely to slip through airport security and, Larkin hopes, 'would open smugglers to terrorism charges on the grounds they were smuggling radioactive substances'.

It's early days but a brilliant idea. Larkin told me, 'You can sum anti-poaching efforts up in three words: guards, guns, and gates.' But Rhisotopes is different. 'The radiation would simultaneously decrease demand *and* put pressure on the smuggling routes.'

Larkin and his team are busy running calculations to see if the radiation will bring any harm to rhinos and their keepers. 'We're currently modelling the radio-sensitivity of a rhino's head to cobalt-60,' Larkin said. Larkin says the radiation is highly unlikely to have any negative impact on the rhino's health, but they need to prove that before they implant entire herds. People would be safe from the horns, too (unless they ground one up and ate it or had it in their suitcase). Besides, as Larkin put it to me, 'the small radiation dose to the rhino is better than the alternative.'

In June 2024, the Rhisotopes team implanted radioactive pins into the first 20 rhinos. The rhinos are being closely monitored to spot the signs of radiation sickness (an incredibly unlikely outcome). Perhaps one day, rhinos invisibly glowing in gamma rays will be commonplace across the African savanna.

I asked Larkin what comes after rhinos. He answered immediately. 'Pangolins. They're the most trafficked animal in the world. We want to use this technology to save as many animals as possible.' It won't be easy to spike the dog-sized critters with cobalt-60, because they're covered in scales and lack the convenience of a protruding horn. But 'we're going to take some high-resolution X-ray CT scans of a pangolin soon to see where on its body we might stash the radioactive pin.' Demand for their meat and scales in Eastern black markets threatens the pangolin with extinction.[38] Cobalt-60 might be the most effective tool we have of saving them.

Before I could even ask the question, Larkin had read my mind: 'We've got our eyes on elephants, too. It might be possible to make their tusks radioactive and put a dent in the ivory trade.' Elephants will be trickier than rhinos; whereas rhino horn is toenail-like, elephant horn is tooth-like. But, as with the pangolin, it won't stop Rhisotopes from trying.

In 1992, Knipling and Bushland won the World Food Prize for 'sustaining vast sources of food, especially livestock and wildlife populations, and consequently ensuring human health' using their sterile insect technique. Bushland passed away in 1995, and Knipling 5 years later. They were both posthumously awarded the 2016 Golden Goose Award for their 'silly-sounding science that's returned serious benefits to society'.[39]

Chapter 11. Radioactive Remedies

Now is the time to understand more — so that we may fear less.

– Glenn Seaborg, 1969

In the evening on Friday, 8 November 1895, German physicist Wilhelm Röntgen was working alone in his darkened lab. He was passing powerful electric currents through a *Crookes tube* — a glass tube out of which almost all the air has been pumped. The beam of electrons glowed as they zapped through the tube's tenuous air.

Röntgen had encased his electrified Crookes tube in thick cardboard to stop it shining into his lab. He was about to turn off the current, but something caught his attention. A green light shimmered in the darkness on the other side of the room. Röntgen excitedly struck a match to illuminate its source. The scintillations were coming from a screen daubed in fluorescent salts. It was impossible, surely; the light from the Crookes tube was being completely blocked by the cardboard, and electric currents don't travel that far through air. But Röntgen's eyes were not deceiving him. The salts were aglow.[1]

Nobody before had seen anything of the sort. Whatever was responsible for the fluorescence was radiating invisibly from Röntgen's Crookes tube. He named the spectre *X-rays*, 'for the sake of brevity'. Many of Röntgen's contemporaries suggested calling them *Röntgen rays* in his honour. As a modest and introverted man, he protested, but the name persists in more than a dozen languages including Dutch (*röntgenstralen*), Swedish (*röntgenstrålar*), and Polish (*promienie rentgenowskie*). Element number 111 — *roentgenium* ('Rg') – is named in his honour.

We know today that the current in Röntgen's Crookes tube was stimulating the atoms of the glass, causing them to shine – but invisibly to us. Human eyes cannot perceive X-rays directly because they lie far beyond violet hues at the upper end of the visible spectrum. Röntgen therefore made X-rays *indirectly* visible by shining them onto fluorescent salts. The salts, dancing with the energy of the incident radiation, twinkled in his dusky lab.

Intrigued, Röntgen experimented with placing paper between his Crookes tube and the shining salts. They glowed just as brightly. The X-rays were passing *through* the paper, as though it wasn't there. A deck of playing cards wasn't enough to dim their brilliance. Nor was a book of a thousand pages. He had more luck with sheets of wood, though, which dulled them a little. Sheets of metal attenuated them even more.

Next, Röntgen tried something else. He rested objects on photographic plates and exposed them to the invisible rays. He tried a bobbin wound by a wire, a set of weights, and a compass. On developing the photographs, the shadow of each object appeared as a silhouette.

And then Röntgen tried something legendary. He rested the left hand of his wife, Anna Bertha Ludwig, on a photographic plate and exposed it to the invisible rays. On developing the photograph, the shadows of her long finger bones faded into view. X-rays shone through flesh effortlessly, *but not through bone!* It was the world's first image of the inside of a person taken from the outside.[2]

Ludwig apparently exclaimed *ich habe meinen Tod gesehen!* – 'I have seen my death!' – when she saw her skeleton fingers. For his discovery, Röntgen became the first ever recipient of the Nobel Prize in Physics in 1901.

The popularity of Crookes tubes meant other scientists could easily duplicate Röntgen's experiments and make X-rays of their own. Their medical applications became immediately obvious. Within a year of Röntgen's discovery, X-rays had left the lab and entered the hospital. Thus began the longstanding practice of using radiation as a diagnostic tool.

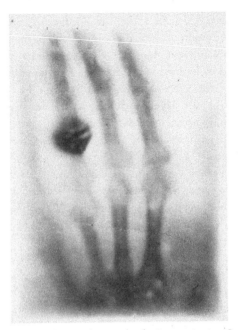

The first X-ray of human bones, taken by Röntgen in 1895. Ludwig was wearing a ring on her third finger. Image by W. K. Röntgen, courtesy of the Wellcome Collection.

Seeing with antimatter

We spend a lot of time and money shielding people from radiation. A considerable clutch of the rules of radiation protection – implemented in all nuclear facilities – are designed to stop nuclear materials getting inside people. You might be surprised to learn, then, that there are procedures in which physicians intentionally inject radioactive substances into patients.

One of the most widespread nuclear diagnostic tools is *positron emission tomography*, more commonly known as a PET scan. You might know somebody who's had one. You might have had one yourself. It's used to search for hard-to-spot diseases such as Parkinson's and epilepsy, but it's mostly used to find cancerous tumours.

It's an elegant imaging technique that makes X-rays look luddite by comparison.

Each subatomic particle – be it a proton, neutron, or an electron – has an equal and opposite counterpart; equal in the sense that they have the same mass, but opposite in the sense that other properties, such as electrical charge, are reversed. We call these mirror-image counterparts *antimatter*. The proton's antimatter counterpart is (imaginatively) named the antiproton. The neutron's is the antineutron. And the electron's is (in defiance of predictable naming schemes) the *positron*, so called because it carries a positive electric charge. Whereas some radioactive isotopes decay by firing out an electron, others fire out a positron. By analogy with beta decay, we call it *beta-plus decay*, or *positron emission*.

Once it's been fired from an atom's nucleus, a positron isn't long for this world. When matter and antimatter collide, they annihilate immediately. And because the world is made of matter – atoms surrounded by electron clouds – a positron meets its end quickly. Positrons and electrons engage in mutual obliteration, but they leave behind a trace when they vanish. Their combined mass (m) is converted into a pulse of energy (E) according to the invariable exchange rate (c^2) in Einstein's famous equation: $E = mc^2$. Since all electrons and positrons have the same mass, E is entirely predictable when they annihilate: 1,022 kiloelectron-volts, to be exact.* This flash of energy manifests as gamma radiation.

When an electron and positron collide, they create not one but *two* gamma rays. The energy is shared equally between the pair, giving rise to twins that each carry exactly 511 kiloelectron-volts. If you detect a 511-kiloelectron-volt flash on your gamma-ray detector, you can be sure that it's an echo of a dying positron.

And there's one more twist in this tale of mutual annihilation. The gamma-ray twins shoot away from the blast's epicentre in exactly opposite directions. And that's useful to know. Because if we detect

---

* Electrons and positrons have identical masses – m – of 9.11×10^{-31} kilograms. Plugging our numbers into $E = mc^2 - 9.11 \times 10^{-31} \times (299{,}792{,}458)^2$ – yields 8.19×10^{-14} joules of energy. Multiply that by 2 (because there's an electron *and* a positron) and then convert your answer from joules to electron-volts.

a pair of 511-kiloelectron-volt gamma rays and, crucially, record the direction whence they came, we can infer the exact spot where they originated.

Which brings us back to those cancerous tumours. By making cancer radioactive – specifically, by making it glow with 511-kiloelectron-volt gamma rays – physicians can pinpoint its whereabouts inside a patient without so much as wielding a scalpel.

We make tumours radioactive by lacing them with unstable isotopes. Choice of isotope is key. For a PET scan, it must decay *via* positron emission. And, critically, it must have the right half-life. Too long, and the gamma rays will shine feebly; too short, and the atoms decay into nothingness before they reach the tumour. The half-life must fall in a radioactive Goldilocks zone, somewhere between a few hours and a few months. Given that radioactive half-lives vary from millionths of a second to many billions of times the current age of the universe, the Goldilocks zone is sparsely populated.

A few positron-emitting isotopes fall into the radioactive Goldilocks zone, but the one used most often in PET scans is the rhythmic-sounding fluorine-18. With its 1 hour and 50 minute half-life, it leaves just enough time to get it from a test tube in the lab into a syringe on a hospital ward before it decays away.[3]

Now, physicians can't just intravenously administer fluorine-18 in its pure form. Once inside the body, it would drift aimlessly, and our gamma-ray detectors would see nothing but a blurry person-shaped glow. They have to concentrate the fluorine-18 inside tumours. This is where nuclear chemists play their part. Like confectioners, they attach the fluorine-18 atoms to molecules that are irresistible to cancer cells. Tumours can't help but gobble them up as they pass by in the bloodstream.

So, nuclear chemists bake fluorine-18 atoms into a molecule called *fluorodeoxyglucose*. You probably haven't encountered fluorodeoxyglucose, but you definitely *will* have encountered a similar molecule: glucose, a staple of everyday life. It turns out that many cancers love glucose. They lap it up. And in their greed, tumours often fail to discern between fluorodeoxyglucose and the real thing.

Physicians inject the radioactive sugar into their patient's veins.

Sketches of humdrum glucose (left) and radioactive fluorodeoxyglucose (right). Note how similar their molecular structures are.

Then, with an undiscerning pallet, cancerous tumours inadvertently make themselves radioactive. Inside the tumours, fluorine-18 atoms fire out a stream of positrons, which engage in mutual annihilation with nearby electrons. By placing the patient inside the PET scanner – a ring of gamma-ray detectors – physicians can pinpoint the tumour: when two opposite-facing detectors simultaneously register a 511-kiloelectron-volt gamma ray, it follows that a tumour must lie somewhere between them. By detecting gamma-ray pairs from multiple angles, the location of the cancer can be triangulated with pinpoint accuracy.

Modern PET scanners can reconstruct the shape of cancers in three dimensions. And, since tumours absorb the radioactive sugar no matter where they are in the body, PET scans even shed light on tumours nobody knew were there.

Nowadays, gamma-ray detectors are so sensitive that less than a billionth of a gram of fluorine-18 is enough for a detailed PET scan. After 12 hours, 99 per cent decays into non-radioactive oxygen-18. Some of that energy, inevitably, ionises atoms in the patient's body and gives them a radiation dose. But it's nothing to worry about. A PET imparts something like 8 millisieverts to a patient, which is about the same as somebody living in Cornwall receives naturally every year.[4]

And there are no side effects ... other than the discomfort of having to lie still for an hour whilst the detectors whirr away.

Radiotherapy

Cancer kills 27,000 people every day and is on the rise globally. It represents one of the twenty-first century's most profound human and economic problems.[5] And it's difficult to cure, partly because it's not really a single disease at all – it's many hundreds of them. Whilst its manifestations are manifold, the underlying cause is always the same: unrestrained cell division. Those rampant cells overwhelm the body as they grow into tumours or, in the case of leukaemia, destroy the immune system. Radiation (in huge amounts) causes cancer. But paradoxically, it can also be one of cancer's most effective cures.

Physicians were attempting to cure cancer using X-rays within a year of Röntgen's discovery. One of the earliest recorded efforts was made by a Frenchman named Victor Despeignes. In 1896, he discovered a tumour the size of a large orange whilst examining a patient complaining of stomach-ache. He prescribed plenty of milk, the occasional swig of medicinal wine, and a course of X-rays from a Crookes tube to the abdomen. Early radiotherapy techniques were crude, but the principle endures: by exposing tumours to bright radiation spotlights, we slow the growth of malignant cells or kill them altogether. Physicians nowadays tend to use gamma radiation because its high energy carries it deep into the body, thereby reaching tumours that lower-energy X-rays never could.

Modern radiotherapy is much more clinical, too. Gamma rays are normally generated by smashing electrons into a tungsten target at near light speed: the atoms in the target, stimulated by the electric beam, shine brightly in gamma rays. The rays shine in all directions as from a lightbulb, and so they're clipped into pin tight spotlights that match the shape of the tumour.

For the most delicate of radiotherapeutic procedures – like those designed to blast brain tumours – physicians wield so-called *gamma knives*. But these knives have no blade. They're a series of tightly focused gamma beams – some 200 in total – that shine into a patient's brain from individual lumps of radioactive cobalt-60. The beams enter the patient's head from different angles and intersect at the tumour, thus maximising radiation damage to cancerous cells whilst minimising collateral damage to healthy brain tissue. It's an alternative to open brain surgery, which can be risky if the tumour is embedded deep inside the patient's skull. Gamma knives carry zero risk of infection and, despite their name, are painless.

Every year, more than 6 million patients have their cancers blasted by radiation beams. Access to radiotherapy across the world, however, is uneven. The UN Scientific Committee on the Effects of Atomic Radiation estimate that more than 90 per cent of the radiotherapy courses are administered to the richest half of the global population. This will change as the developing world becomes richer. In the meantime, the International Atomic Energy Agency supports developing nations in setting up radiotherapy treatment centres through its Rays of Hope initiative.[6]

Radioactive thyroid

Radiotherapists blast cancerous tumours with radiation from *outside* the body. But there's another medical speciality concerned with irradiating tumours from the *inside*. These medical marvels are called *nuclear medicines*, and they rely not on radiation *beams* but on radioactive *atoms*.

Radioactive iodine-131 – a common fission fragment in spent fuel, which caused thyroid cancers in the wake of Chernobyl – can be used as a *cure* for thyroid cancer.[7] Most people with thyroid cancer often have part of their thyroid surgically removed. Some even have the whole thing cut out. The surgery is usually successful, but it can be difficult to know for sure whether all the cancer has been excised. In these cases, patients are sometimes given a dose of iodine-131. It fires out beta particles with a half-life of 8 days, placing it in the radioactive Goldilocks zone. The radioactive payload swarm through the bloodstream, homes in on the thyroid, and destroys the remaining malignant cells.[8]

The thyroid-seeking iodine locks on to remaining thyroid cells and largely ignores everything else, meaning there's minimal collateral radiation damage to healthy cells. It also sweeps away malignant thyroid cells that have spread to other parts of the body. It's precision medicine. A few hundred billionths of a gram or so of iodine-131 usually does the trick, which doesn't sound like much, but the tiny arsenal contains about a hundred times more atoms than there are cells in the human body.[9]

Ingesting iodine-131 inevitably makes patients more radioactive than the average person on the street. Immediately after she's taken an iodine-131 pill, a patient will be even more radioactive than most of the samples in my lab, but her radioactivity diminishes swiftly. (Not to mention, she'll excrete it through normal bodily functions.) It's all but gone after a couple of months. In the meantime, though, elevated radioactivity can cause a minor nuisance; patients occasionally set off radiation alarms in airports days or even weeks after treatment.[10]

Guidelines vary from hospital to hospital, but iodine-131 patients are generally advised to stay away from people for a day or two after therapy. Such advice is probably overly conservative, though. A study by researchers at Washington University in the year 2000 tracked 30 patients over the 10 days following treatment. Patients slept alone and avoided close contact with others for 2 days, after which point most of their radioactivity had diminished. Researchers rigged the patient's houses, loved ones, and pets with dosimeters (similar to the ones I wear at work) and found average exposures were 0.24 millisieverts, or 5 weeks' worth of background radiation. The pets were exposed to slightly larger doses than the human members of the households; I wonder if the patients couldn't help but give their cat a cuddle when they returned home from the hospital.[11]

Aside from the small feat of curing cancer, iodine-131 is also used to remedy overactive thyroids. By dosing the patient with just the right amount of radioactive iodine, physicians partly destroy it and curtail its overzealousness.[12]

Cancer-seeking molecular machines

Treating thyroid ailments with radioactive iodine is relatively straightforward because the human body guides the therapeutic isotope – iodine-131 – to the right place with minimal medicinal persuasion.

The same is true for strontium-89, a Goldilocks beta-emitter with a 7-week half-life. Strontium has similar chemical characteristics to calcium and therefore has an affinity for bonestuff, earning its nickname as a 'bone-seeker'. Thus, strontium-89 atoms administered intravenously target malignant bone cells. Not all nuclear medicines are curative like iodine-131; some, like strontium-89, are palliative, relieving pain during the final months of the patient's life.[13]

There's a cornucopia of other radioactive isotopes in the Goldilocks zone that could, in principle, be used to fight cancer, but guiding them to exactly where they're needed is a biochemical conundrum. Take lutetium-177, for example, a beta-emitter with a 6.5 day half-life that on paper is a medicinal wonder. However, neat lutetium-177 doesn't go anywhere in particular once it's inside the body, and indiscriminately exposes all cells to its radiation as it courses aimlessly. It's like a radioactive scattergun.

Nuclear chemists, therefore, engineer molecular machines that deliver lutetium-177 to cancerous cells whilst steering them away from healthy ones. These cancer-seekers have begun to be deployed in the treatment of prostate cancer.

Prostate cancer is the second-most prevalent cancer in the world, second only to breast cancer. It kills more than 350,000 men annually. The good news is that if it's caught early, 10-year survival rates can be as high as 99 per cent; prostate tumours are simply removed in surgery or obliterated by radiotherapy. The bad news is that if it spreads to other parts of the body, the chance of surviving long-term plummets. Late-stage prostate cancer can be managed for a while using hormone therapy and chemotherapy, but it often becomes resistant to treatment. It's frequently fatal. Nuclear medicine is making strides to alleviate suffering in such situations.[14]

Healthy prostate cells usually coat themselves lightly in a protein called *prostate-specific membrane antigen*, or (more conveniently) *PSMA*. Cancerous prostate cells take it to the extreme by daubing themselves in it. With garish PSMA-shaped targets on their backs, malignant cells stand out against the backdrop of healthy ones.

In 2015, researchers in Germany designed a PSMA-seeking molecule called *lutetium vipivotide tetraxetan*, normally shortened to *Lu-PSMA-617*. The molecule, assembled from building blocks of carbon, hydrogen, nitrogen, and oxygen, features a pair of grabbers: one grabber holds a lutetium-177 atom and the other locks onto PSMA targets. The entire molecule – assembled from 143 atoms – is a few nanometres across. You could fit 33,000 of them side by side on a strand of human hair.

When it's injected into the bloodstream, live Lu-PSMA-617 homes in on malignant prostate cells and delivers unto them a lethal dose of radiation. It finds them regardless of where they are in the body. It's analogous to the glucose-like molecule that delivers fluorine-18 during a PET scan, only it wasn't designed to simply illuminate cancerous cells: it was designed to destroy them.[15]

Starting in 2018, 551 men with late-stage prostate cancer were given bouts of experimental lutetium-177 therapy in a randomised clinical trial. Those treated with lutetium-177 lived for 8.7 months, on average, without their cancer getting worse; those treated with regular end-of-life care lived for 3.4 months, on average, without their cancer getting worse. Lutetium-177 also extended their lives by an average of 4 months. A few millionths of a gram of lutetium-177 is all it took for an effective dose.[16]

Four months of extra life is a lot to somebody with late-stage cancer. Scaling those extra months across the 350,000 men who die annually of prostate cancer equates to 117,000 extra years of life. Small gains on the individual level yield enormous benefits to society.

Lutetium-177 therapy was approved for end-of-life care by regulators in the USA in March 2022 and in the EU nine months later. Lutetium-177 has been described as 'one of the greatest success stories in the history of nuclear medicine'. It required the combined efforts of biologists, chemists, biochemists, pharmacists, oncologists, medical practitioners of all walks, radiation physicists, nuclear chemists, and – of course – hundreds of terminally ill men willing to make themselves test subjects.[17]

Therapeutic lutetium-177 is still being finessed. Thirteen per cent of the men in the clinical trial had to have blood transfusions because of collateral damage to their bone marrow. Nausea and fatigue were common. But work is underway to use lutetium-177 in earlier stages of prostate cancers. In 2023, there were at least 20 new lutetium-177 drugs in various stages of development worldwide.[18]

Meanwhile, researchers are rifling through the Goldilocks zone for other therapeutic isotopes. Terbium-161 is being researched as another treatment for prostate cancer. Rhenium-188 is being turned into a cure for skin cancer; scandium-47 is being developed as a treatment for ovarian cancer; holmium-166 for liver cancer. Yttrium-90 is already being used to treat liver cancer.* [19]

Despite its success, beta therapy has some inherent drawbacks. Beta particles pass through matter easily; in the context of nuclear medicine, this effectively gives them a large blast radius, and so they often overshoot tumours and damage healthy tissues. They also carry relatively small amounts of energy; it takes multiple hits to kill a malignant cell. No amount of molecular engineering can fix these design flaws. Wouldn't it be convenient if we had a type of radiation that doesn't travel far through matter and carries with it enormous energy? Such radiation exists: alpha.

---

* Yttrium-90 therapy works slightly differently to lutetium-177 therapy. Instead of delivering the radioactivity aboard a cancer-seeking molecule, physicians deliver it inside microplastic beads. The radioactive beads are injected into the tumour's blood supply, where they get lodged in narrow capillaries.

Targeted alpha therapy

Whereas zippy beta particles typically traverse a couple of millimetres through human tissue, alpha particles have a limited range; they usually travel no more than a twentieth of a millimetre – less than a hair's width – before coming to a standstill. Alpha particles are far more destructive, too. They're 7,000 times more massive than beta particles, and carry far more energy. If beta particles are bullets, then alpha particles are cannonballs.[20]

Alpha's small blast radius and intense potency means it delivers enormous amounts of energy close to the atom whence it came. If alpha-emitting isotopes are delivered to the right part of the body – a cancer cell, for instance – they obliterate their targets with a few hits, whilst inflicting hardly any collateral damage. And therein lies the rationale behind *targeted alpha therapy*. The one-hit wonders are the superstars of nuclear medicine.

To date, only one has been approved by the FDA: radium-223, which alpha decays with a half-life of 11.5 days. It's sold and administered under the name Xofigo as a treatment for metastatic bone cancer. Like strontium, radium is a bone-seeker. Thus, Xofigo administered intravenously delivers radium-223 to bone metastases. Whilst strontium-89 is purely painkilling, radium-223 is painkilling *and* curative, prolonging survival time and delaying cancer progression.[21]

Radium-223 is just the start. There's a whole alpha-emitting medley in the Goldilocks zone that could be turned into medicine, but like fluorine-18 and lutetium-177, they must be bolted into molecular machines.

Actinium-225, with a 10-day half-life, is one such isotope. It holds legendary status amongst nuclear scientists because it goes through *four* successive alpha decays as it cascades rapidly down the periodic table towards non-radioactive bismuth-209.* With

---

*There are two beta decays in there, too. The full decay chain goes: actinium-225 → francium-221 → astatine-217 → bismuth-213 → polonium-213 → lead-209 →

each successive decay, another malignant cell is demolished, meaning tiny amounts of actinium-225 can devastate large amounts of cancer. It's like a cannon that automatically reloads before firing again, and again, and again.[22]

The repeat rounds of alpha decay are a mixed blessing, though. Just as real cannons lurch backwards from the force of firing a cannonball, atoms kick back from the force of an ejected alpha particle. Actinium-225 and its progeny sometimes recoil with enough force to break free from their molecular machine. Detached from the cancer-seeker, they carry their radioactivity to other parts of the body and damage healthy tissues. 'Actinium-225 is either the holy grail or a bloody nightmare, depending on whom you ask,' a colleague once told me.

Work is ongoing to design molecules that cling on to alpha-emitters more tightly when injected, and initial experiments are encouraging. Clinical trials conducted between 2016 and 2023 gave 488 men with late-stage prostate cancer doses of actinium-225 at 8-week intervals. Conventional treatment options had run out: they'd previously been given chemotherapy, bouts of lutetium-177, rounds of radium-223, and a cocktail of other drugs. The results were astonishing: median survival time without their condition worsening was 7.9 months, none of them had serious side effects, and none of them died from the treatment. For now, actinium-225 lies squarely in the early stages of experimental medicine, but it's progressing towards deployment in hospitals.[23]

Other trials are in the early stages of turning actinium-225 into cures for certain leukaemias and cancers of the colon, eye, and white blood cells.[25]

Meanwhile, alpha-emitting astatine-211 is in the early clinical stages to become a cure for thyroid cancer; polonium-212 and bismuth-212 for lung cancer and neuroendocrine tumours; bismuth-213 for leukaemia; and thorium-227 for lymphoma and ovarian cancer. As well as curing cancer, researchers are also turning

---

bismuth-209. You can tell the alpha decays from the beta decays because the former involve a mass number decrease of 4.

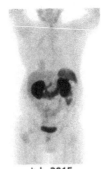
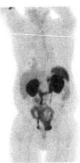

December 2014
Pre-treatment

July 2015
After three bouts of treatment

September 2015
After a fourth bout of treatment

PET/CT scans of a patient from a clinical trial in 2014–2015 during repeat rounds of actinium-225 therapy.[24] Image courtesy of Society of Nuclear Medicine and Molecular Imaging.

the arsenal towards infectious diseases such as HIV, meningitis, and sepsis.[26]

Note the words 'early stages' appear frequently in the context of these isotopes. Targeted alpha therapies are new medicines, and it will probably be decades before they're implemented widely, but it's reassuring to know that some of the brightest minds go to work every day to make it happen.

You can't just walk into your local chemist and buy a few crumbs of fluorine-18 over the counter. That's because, aside from the complications that come with it being incredibly radioactive, the fluorine-18 used in PET scans doesn't exist naturally. Nor does the iodine-131 used to treat malignant thyroids. Nor does strontium-89, lutetium-177, radium-223, actinium-225, or most of the other medicinal isotopes. Their absence in nature makes medicinal isotopes some of the rarest substances on the planet.

So, how do we get our hands on them? We have three options: by smashing atoms, by generating neutrons, or by milking cows.

Atom smashers

By 1925, Rutherford and Blackett sparked the first human-made nuclear reactions and alchemised oxygen by bombarding nitrogen atoms with alpha particles. Following their lead, Irène Joliot-Curie – Marie and Pierre Curie's daughter – and her husband, Frédéric Joliot-Curie, did a similar thing in 1934. At the Institut du Radium in Paris, they irradiated everyday, non-radioactive aluminium with an intense flux of alpha particles to produce showers of neutrons. But they noticed something strange upon terminating their experiment: their aluminium danced with positrons after the alpha particles ceased to stream. *They'd made their aluminium radioactive!* Or more precisely, as their subsequent chemical tests revealed, they'd alchemised non-radioactive aluminium (13 protons + 14 neutrons) into radioactive phosphorus-30 (15 protons + 15 neutrons). They'd created the first human-made radioactive substance.[27] Their nuclear reaction looked like this:

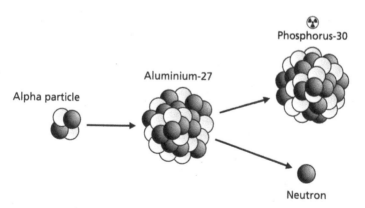

Protons are white; neutrons are black.

Their radioactive phosphorus-30 then positron decayed, with a half-life of 2.5 minutes, into non-radioactive silicon-30. The couple were awarded the Nobel Prize in Chemistry 1935 and thus began a craze within the scientific community for alchemising synthetic isotopes by bombarding targets with radiation.

But alchemising isotopes on lab bench tops is, well ... laborious. In his inaugural Nobel Lecture, Monsieur Joliot-Curie noted 'the weights of elements formed at the present time are less than 10^{-15} grams' which represented 'at the most a few million atoms'.[28] The few million electron-volts carried by alpha particles could scarcely push them beyond the force of electrical repulsion surrounding the atom's nucleus. Scientists were also limited by the relatively feeble amount of radiation that emanates from natural substances. To expand their synthetic chemistry set of synthetic isotopes, they needed something with a little more oomph. Rutherford himself expressed this in his Presidential Address to the Royal Society of 1927:

> It has long been my ambition to have available for study a copious supply of atoms and electrons which have an individual energy far transcending that of the alpha and beta particles from radioactive bodies.[29]

Rutherford's impassioned call was answered by a South Dakotan physicist called Ernest Lawrence with his invention in 1929 of the *cyclotron*: a machine that generates bright beams of artificial radiation that can be directed towards targets. It consists of two semicircular, hollow electrodes called 'dees' (because they're D-shaped), separated slightly but with their straight edges lined up, so they resemble a disc with a channel cut out through the middle. A cloud of charged particles – protons, say – is generated in the gap between the dees, and the dees are electrified with a rapidly oscillating high voltage. This voltage causes the protons to move at higher and higher speeds.

Normally, the protons would accelerate back and forth between the dees along a straight line, but there is another piece of apparatus that bends their path: the dees are sandwiched by two electromagnets, one just above and one just below, which creates a powerful

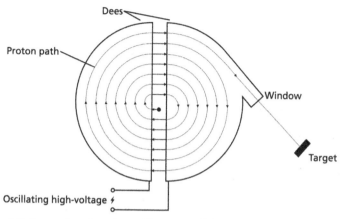

A bird's-eye-view of a cyclotron. The magnetic field rises up through the dees from underneath and bends the ions into a spiral path.

magnetic field running perpendicular to the protons. In the magnetic crosswind, the protons curve in a semicircular arc through one dee, and then back through the other. But because the protons are accelerating, they do not complete a circle but instead move in larger and larger orbits, tracing a path that spirals outwards towards the dees' circumference.

The whole contraption is sealed in a vacuum-tight box to eliminate air resistance and let the protons accelerate to incredible speeds as they whizz through, in Lawrence's words, the 'proton merry-go-round'. A window at the dees' edge allows passage to the stream of energised protons, where they can smash into the nuclei of atoms in a target.[30]

In the autumn of 1930, Lawrence and his PhD student, Stanley Livingston, succeeded in building a prototype. It was the size of a small frying pan and accelerated charged particles to 80,000 electron-volts. A year later, they'd built one the size of a steering wheel that accelerated them to 1.2 million electron-volts. By 1939, the team had expanded, and they'd built a cyclotron with a magnet weighing 220 tonnes and dees 60 inches across, roomy enough to contain several physicists. It delivered 600 trillion charged particles

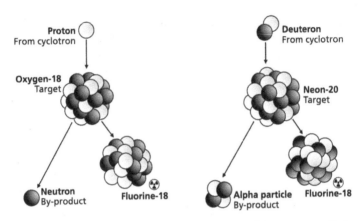

The main production routes for fluorine-18. Protons are white; neutrons are black.

every second with energies of 16 million electron-volts. With such powerful machines, they could accelerate all manner of subatomic particles – alphas, electrons, protons – and smash them into the nuclei of targets from across the periodic table, knocking pieces off to make new isotopes. Lawrence won the Nobel Prize in Physics 1939.[31]

Using the 60-inch cyclotron, McMillan would create the first transuranic element, neptunium, and Seaborg would create the second, plutonium. Seaborg and his colleagues would go on to discover more than 100 new radioactive isotopes, many of them today used in nuclear medicine. One of them was thyroid-seeking iodine-131, which they alchemised in 1938 by smashing deuterons – hydrogen-2 nuclei – into a target of element number 52, tellurium. A decade later, Seaborg's mother developed an overactive thyroid but made a full recovery after a course of iodine-131 therapy.[32]

Today, there are more than 1,500 cyclotrons dotted across at least 95 nations. They're tools of scientific research and the pillars of medicinal isotope production. Some hospitals have their own cyclotrons in the basement.[33]

Take the fluorine-18 used in PET scans, for instance. Physicists usually make it in one of two ways: smash high-energy deuterons (hydrogen-2 nuclei) into neon, or high-energy protons into oxygen-18. The cyclotron energy of the incoming particles blasts pieces away, alchemising fluorine-18 in the process. The fluorine-18 is quickly siphoned off to form medicinal fluorodeoxyglucose.[34]

Astatine-211 is made similarly, by smashing high-energy alpha particles into bismuth targets. And the holy grail of targeted alpha therapy, actinium-225, can be alchemised by walloping radium-226 with protons.[35]

Neutron generators

It's not just cyclotrons that we use to alchemise medicinal isotopes. We use nuclear reactors, too. These so-called *research reactors* are typically far less powerful than those in power stations. Their combined output — all 250 globally — is about the same as a single pressurised water reactor. But research reactors aren't built to generate torrents of electricity. They're built specifically to generate intense fluxes of *neutrons*.[36]

Take lutetium-177, for example. There's no way to make it in sufficient quantities by irradiating a target with protons, or deuterons, or any other particle accelerated in a cyclotron. It can only be made using neutrons, either directly by irradiating lutetium-176 targets or indirectly by irradiating ytterbium-176 targets.[37]

Iodine-131 — the mainstay of thyroid treatment — is made by

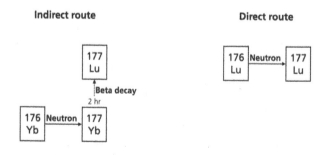

similar means. By irradiating a tellurium-130 target with neutrons, the atoms become tellurium-131, which transform into iodine-131 when they swiftly beta decay.[38]

Some three dozen other isotopes are alchemised in research reactors, including many of the ones we've come across already: scandium-47, yttrium-90, terbium-161, holmium-166, and rhenium-188. The higher the neutrons flux, the better, which is why research reactors run on highly enriched uranium, typically 20 per cent uranium-235 and upwards.[39]

Milking the cow

Much of the actinium-225 in the world comes from a single small lab at Oak Ridge National Laboratory, Tennessee. It's not alchemised in a cyclotron, nor in a research reactor. It's milked from a *thorium cow*.

Between the mid-1940s and mid-1960s, the USA made small amounts of uranium-233 in breeder reactors for research and development on new types of atom bombs. It was never deployed in atomic warfare. Interest in uranium-233 weapons waned, and the focus shifted to using it as the third reactor fuel, leading to the breeding of substantial quantities between 1965 and 1970. But it was never deployed in power stations, either. The uranium-233 was placed into storage, where it will remain until a long-term Onkalo-style solution is devised.[40]

Whilst they haven't been used to generate electricity, those uranium-233 stockpiles haven't been idle. They've been cascading, spontaneously, down a series of radioactive decays towards non-radioactive bismuth-209. The fourth isotope in that decay chain is actinium-225. That makes *old* stockpiles of uranium-233 rich in the most coveted isotope in nuclear medicine. I emphasise 'old' because it's crucial that enough time has passed to allow the uranium-233 progeny's to accumulate.

Unfortunately, you cannot use the actinium-225 in old uranium-233 stockpiles directly because it's mixed with all the other isotopes.

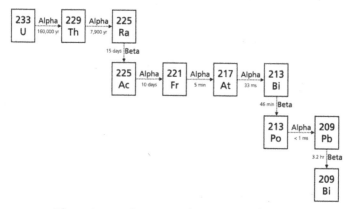

The uranium-233 decay series. The times next to the decay arrows are the half-lives.

Injecting such an assorted array of radioactive isotopes into somebody's bloodstream would be insane. To be used in nuclear medicine, actinium-225 must be separated and purified. This presents challenges on several fronts. Firstly, actinium-225 exists in minute quantities because uranium-233 decays at such a ponderous pace, and separating tiny amounts of one element from vast amounts of another is exceedingly difficult; even after 60 years in storage, 9,997 in every 10,000 atoms are still uranium-233. Secondly, the sample hums with radiation, which means the finicky procedure must be performed using robot arms through thick concrete walls and leaded glass. And finally, even if you did go through the Herculean effort of purifying tiny amounts of actinium-225, its 10-day half-life means it decays into nothingness in just a few months, after which you'd have to do it all over again.

In the early 1990s, radiochemists at Oak Ridge found a neat solution to the problem. They separated and purified the second isotope in the decay chain – thorium-229. It was equal in difficulty to purifying actinium-225, *but they only had to do it once*. Thorium-229 has an almost 8,000-year half-life, some 300,000 times longer than actinium-225's. That means once separated and purified, thorium-229 sticks around for millennia. There's no need to separate more.[41]

Thus, the thorium cow was born. Put those images of black and white bovines from your mind; the thorium cow is simply a sample of thorium-229 run through a chemistry set to separate the purified actinium-225 progeny. An acid solution containing the dissolved mixture is run through an organic substrate in upright test tube open at both ends; the substrate selectively binds to certain elements, teasing them apart such that they drip from the bottom of the test tube one by one. The intense radioactivity emanating from the cow means the procedure must be performed using robot arms.

Like milkmaids in the barn every morning, radiochemists *milk* the thorium cow for its actinium-225 every 2 months or so. The radiochemists really do call this purification procedure 'milking'; it's a technical term. (We should start calling actinium-225 'lactinium'.) And as a dairy cow replenishes its milk through bodily functions, a thorium cow replenishes its actinium-225 through spontaneous radioactive decay. Whilst the thorium-229 is still live – which it will be for millennia – it will produce a steady flow of actinium-225 that can be periodically milked and shipped to research institutions and perhaps, eventually, hospitals.[42]

Uranium-223 stockpiles are generally considered to be nuclear waste. Now they're being repurposed to find new cures for cancer, serving as a reminder that we ought to be careful about throwing away strange nuclear materials that might someday prove useful.

Elsewhere, radiochemists milk different breeds of cow for different isotopes. They include yttrium-90, holmium-166, rhenium-188, lead-212, bismuth-213, radium-223, and thorium-227. The cows are often small, meaning they can fit easily into hospital basements, shortening the time it takes to get from lab to cell.[43]

Unsteady supply

By their nature, radioactive isotopes decay over time, and so they can't be stored for future use. Leave lutetium-177 on the shelf for a few months and it will have turned into hafnium-177; leave actinium-225 unused, and it will decay rapidly into bismuth-209. Medical isotopes must be produced continuously to meet demand, arriving at hospitals on a just-in-time basis. Reliable supply chains are essential but, alas, seldom exist: few facilities have the instruments, feedstocks, or expertise to create them. Their production is concentrated in the hands of an alarmingly small number of suppliers, exposing them to vulnerabilities.

The shaky supply chains occasionally come tumbling down. In 2010, two of the world's most bountiful research reactors – the National Research Universal in Canada and the High Flux Reactor in the Netherlands – were taken offline for repairs simultaneously, disrupting the supply of several medical isotopes. Diagnostic imaging and cancer treatments were delayed as direct consequences, and many treatment centres struggled to give their patients proper care. The High Flux Reactor – a single research reactor – supports more than 60 per cent of all Europe's medical diagnostic procedures. And, along with many of the world's research reactors, it's approaching the end of its operational lifetime.[44]

Medicinal cows are also rare beasts. For instance, only three nations milk appreciable quantities of actinium-225 from thorium cows: the USA, Germany, and Russia. Their rate of production isn't enough to satisfy the world's current actinium-225 demand, and it's still only in the early stages of development. As things stand, a mass rollout to labs and hospitals is impossible, even if it were developed into a cancer-killing wonder drug.[45]

An uninterrupted and plentiful supply of isotopes is perhaps the biggest challenge faced by the nuclear medicine enterprise. Most research papers cite tenuous supplies as a hindrance to treatment: the anxiety of researchers and physicians is palpable.

But the situation is curable. Building more research reactors is an obvious place to start. In 2021, researchers at the Netherlands Cancer Institute concluded, 'Continued investments in high-flux neutron irradiation facilities will be crucial to maintain sufficient availability of lutetium-177 and other medical isotopes'. To that end, the Pallas research reactor is being built to replace the ageing High Flux Reactor. Its foundations were laid in 2022, and when it's finished sometime around 2030, it will supply most of Europe's medical isotopes.[46]

Existing reactors will play a part, too. In 2022, a commercial nuclear power station – a CANDU reactor in Ontario, Canada – produced lutetium-177. The capability was retrofitted into a reactor built in the mid-1980s, long before the therapeutic properties of lutetium-177 were known. It was the first instance of a short-lived medical isotope being alchemised in a commercial power reactor. Perhaps one day, medical isotope production will be built into nuclear power stations as standard.[47]

Researchers in North America, Russia, France, and South Africa are working on new ways of producing isotopes altogether. For instance, as well as milking actinium-225 from thorium cows, it can be alchemised by smashing high-energy protons into natural thorium targets using particle accelerators. This method of production, if implemented fully over the coming decades, could produce thousands of times more actinium-225 every month than all the world's cows combined.[48]

When the Joliot-Curies discovered artificial radioactivity, Irène's mother was ailing from leukaemia. Marie Curie didn't live to see her daughter and son-in-law receive their Nobel Prize, but she did live to see their alchemised phosphorus-30. Monsieur Joliot-Curie would later recall the moment:

> Marie Curie saw our research work and I will never forget the expression of intense joy which came over her when Irène and I showed her the first artificially radioactive element in a little glass tube. I can still see her taking in her fingers (which were already burnt with radium) this little tube containing the radioactive compound ... To verify what we had told her she held it near a Geiger-Müller counter and she could hear the rate meter giving off a great many 'clicks'.[49]

It was indeed a great source of satisfaction for our lamented teacher Marie Curie to have witnessed this lengthening of the list of radioelements, which she had had the glory – in company with Pierre Curie – of beginning.[50]

If only we could travel backwards in time and tell the Curies that a century later, we'd be using their discoveries to cure cancer on the smallest scale possible: one radioactive atom at a time.

Chapter 12. Stockpiles and Sleuths

Panic on the streets of London,
Panic on the streets of Birmingham,
I wonder to myself,
Could life ever be sane again?

– 'Panic', The Smiths, 1986

In September 1945, Julian Webb – a physicist working at Kodak's photography research lab in New York state – was investigating a batch of faulty X-ray film. Numerous customers were finding their developed film partly exposed and covered in a constellation of tiny dark spots, despite never having taken it out of its packaging. Webb suspected the packaging was contaminated with radioactivity. X-ray film is sensitive to X-rays, after all, and so it doesn't take much exposure to leave a mark.

At the time, radiation blemishes contaminated lots of America's paper and cardboard. It came from radium paint, used in war factories to make glow-in-the-dark watch faces and aircraft dashboards. Feeling the squeeze of wartime scarcity, paper and cardboard from these factories was pulped and recycled, and tiny flecks of radioactive paint made their way into the general supply. Paper and cardboard became sporadically contaminated with radium and, if they were used as X-ray film packaging, caused spotty exposures.

Yet Kodak went to great lengths to keep its X-ray film packaging radium-free. They were in total control of raw materials, using only the freshest ingredients. They also relied on specialist paper mills, free from pulp tapped off the general supply chain. And so the dark spots puzzled Webb. Had radium somehow contaminated Kodak's pristine cardboard? Probably not. Webb noticed that spots on adjacent X-ray films *lined up with one another*. The radiation, whatever it was coming from, had shot through several layers of film like a bullet. This suggested it wasn't coming from radium, because radium is an alpha-emitter, and alpha particles are blocked by a single sheet. This radiation was far more penetrating; it was either beta or gamma.

Webb realised that all the spotty film had been in contact with the same batch of packaging, manufactured in a single run at an Indianan paper mill on 6 August. Determinedly, he punched out hundreds of small areas of cardboard that were in direct contact with exposed areas of film and put them in a radiation detector. The complete absence of alpha radiation ruled out radium, and other suspect alpha-emitters such as uranium and thorium, definitively. But Webb did detect lots of beta radiation.

Over the following months, Webb measured the mystery isotope's radiation intensity. Predictably, it waned as the isotope decayed. On plotting the falling intensity on a graph, he realised it was decaying exponentially. *He was measuring its half-life!* Meanwhile, he measured its energy by passing it through aluminium foil of different thicknesses.

Webb was looking at a beta-emitting isotope, with a half-life of about a month and an energy of 600 kiloelectron-volts. Only one isotope has these characteristics: radioactive cerium-141. Where on earth was it coming from? It doesn't exist naturally. Its short half-life meant it must have been alchemised shortly before it contaminated the paper in August 1945, perhaps in June or July, else it would have decayed into nothingness by the time Webb got it into his lab.

The answer was undeniable: cerium-141, as Webb would write, 'is one of the more prolific fission products of the atom bomb'. The cerium-141, along with a swarm of other fission fragments and weapons-grade plutonium, came from the fires of the world's first fission bomb, detonated in the New Mexico desert at 05:29:21 local time on 16 July 1945, more than a thousand miles away from the Indianan mill.[1]

Clouds and craters

The fireball from Gadget – the world's first atom bomb – bloomed and dragged a thick stem of searing smoke in its wake as it reached skyward. The flames illuminated a rolling mushroom cloud, towering at 38,000 feet: the first glimpse of what came to symbolise the awesome destructiveness of the atom. The solemn words of Robert Oppenheimer – director of the Manhattan Project remembered today as the father of the atomic bomb – ring through the annals of history:

> We knew the world would not be the same. A few people laughed, a few people cried, most people were silent. I remembered the line from the Hindu scripture, the Bhagavad Gita. Vishnu is trying to persuade the Prince that he should do his duty, and to impress him, takes on his multi-armed form and says, 'Now, I am become Death, the destroyer of worlds.' I suppose we all thought that, one way or another.[2]

The list of spectators reads like a roll call of the greatest minds in twentieth-century physics. It includes Edwin McMillan, the first to alchemise a transuranic element; Ernest Lawrence, inventor of the cyclotron; James Chadwick, discoverer of the neutron; Isidor Rabi, discoverer of nuclear magnetic resonance, which forms the basis of MRI scanners; Richard Feynman, prolific quantum physicist and science populariser; and Otto Frisch, who seven years previously had explained the theoretical basis for nuclear fission with his aunt, Lise Meitner. Frisch later said it was 'as if somebody had turned the Sun on with a switch' when the bomb exploded.[3]

Forty seconds after the explosion, the shockwave reached Enrico Fermi, who was standing 10 miles from ground zero. It was just 2.5 years since he'd sent the first nuclear reactor critical in Chicago. Cementing his reputation as a legendary experimentalist, he dropped small pieces of paper to the ground as the air blast thundered past. By measuring how far the scraps were blown backwards – about 2.5 metres (from 10 miles away!) – he estimated Gadget's explosive energy. Atomic blasts are so fierce that we need a new unit of energy to quantify them: thousands of tonnes of TNT equivalents, or *kilotons* for short. Fermi estimated Gadget released about 10 kilotons. He'd underestimated; it was actually closer to 20.[4]

Gadget derived its explosive power from a 6-kilogram ball of plutonium-239. Six kilograms of plutonium – a sphere the size of an onion – is all it took to create a mushroom cloud that would dwarf a Himalayan mountain.[5]

There are two main types of fission bomb: implosion-type and gun-type. Step-by-step instructions for manufacturing and assembling each are closely guarded state secrets, but in their basic workings, both generate a blast by sending as much fissile material super-critical as possible in as short a time as possible before the bomb blows itself to smithereens.

An implosion-type atom bomb holds a sphere of plutonium – enriched in fissile plutonium-239 – in its core. The plutonium is sub-critical; stray neutrons, on average, leak from its surface before they strike another atom, dashing any hope of a chain reaction. To send it super-critical, conventional explosives detonate simultaneously around the sphere, and an inward-driving shockwave crushes it like a stress ball. The atoms are squeezed together, closing the gaps through which stray neutrons escape. The sphere becomes super-critical and monstrous quantities of energy are released as a profusion of atoms break apart. Gadget was an implosion-type

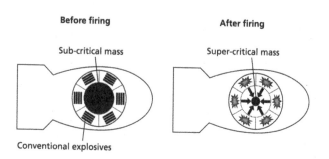

Implosion-type atom bombs compress a sphere of plutonium to super-criticality.

bomb. So was Fat Man, the 21-kiloton bomb dropped on Nagasaki 24 days later.[6]

A gun-type atom bomb is far simpler. A sub-critical mass of uranium – highly enriched in fissile uranium-235 – is fired by conventional explosives at another sub-critical mass. When the two collide, the combined mass goes super-critical. Manhattan Project scientists were so confident it would work that they didn't even test the design before Little Boy, a 15-kiloton bomb, was dropped on Hiroshima just 21 days after Gadget was detonated.[7]

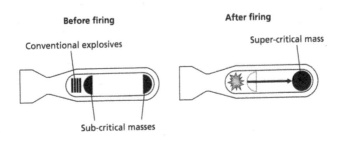

Gun-type atom bombs slam two sub-critical masses together to achieve super-criticality.

Whilst atom bombs are easy enough to conceptualise, they're extraordinarily difficult to make in practice. Miniaturising them so they fit inside long-range missiles that can hit targets hundreds or thousands of miles away – at which point they become useful agents of war – is a formidable challenge. But in August 1949, the Soviet Union managed it and became the second nation to detonate an atom bomb. Thus commenced the atomic arms race. Scientists on both sides of the Iron Curtain quickly found ways to make fission bombs even more powerful (and smaller), and within 10 years of Gadget, explosive yields increased from a few tens of kilotons to many hundreds.[8]

But to push beyond kilotons and into the realm of megatons – *millions* of tonnes of TNT equivalent – a new type of atom bomb was required: a *hydrogen bomb*, which acquires its explosive power from a combination of fission *and* fusion. American scientists detonated the first – codenamed Ivy Mike – on a tropical atoll in the Pacific Ocean in November 1952. The 10.4-megaton blast was 700 times more potent than the fission bomb dropped on Hiroshima. It vaporised the island on which it was detonated and gouged a mile-wide crater beneath its 120,000-foot-high mushroom cloud. Nine years later, Soviet scientists unleashed the most colossal explosion ever wrought by our species when they detonated the hydrogen bomb Tsar Bomba in northern Russia. The 50-megaton blast shattered windows 900 kilometres away, spawned a mushroom cloud seven times the height of Mount Everest, and triggered seismic waves that thrice circumnavigated the Earth.[9]

A hydrogen bomb dropped on a city would kill millions of people immediately. The mushroom firestorm would incinerate many of the survivors. The total collapse of all surrounding infrastructure would finish off many more. An atom bomb is like an earthquake, hurricane, and wildfire all at once.

A war between two nations where a few hundred atom bombs were exchanged in quick succession over urban areas could cool the Earth by up to 5 °C and curtail rainfall severely. A full-scale atomic war would be even more devasting. Tens of millions would die in the blasts. Hundreds of millions, or even *billions*, might die as nuclear winter envelops the Earth and cripples the world's food supply. The global economy would crumble, and our institutions would be thrown into turmoil. It would be like us dropping an asteroid on ourselves.[10]

Our 200,000-year human history would be split into a 'before' and an 'after'; civilisation as we know it would be replaced by something post-apocalyptic. A world in which *thousands* of atom bombs were exchanged is a threat that makes all others – pandemics, artificial intelligence, climate change – look trivial. It's existential. Carl Sagan summarised the situation in 1983:

> Imagine a room awash in gasoline, and there are two implacable enemies in that room. One of them has 9,000 matches. The other has

7,000 matches. Each of them is concerned about who's ahead, who's stronger ... What is necessary is to reduce the matches and clean up the gasoline.[11]

It's a bleak picture. But let's look at some numbers.

Fallout

In the twilight years of the 1940s, there were six atom bomb tests. In the first half of the 1950s, there were 63; in the latter half, there were 228. The USA and the USSR conducted more than 90 per cent of those tests between them, partly in the name of scientific inquiry – there's no better way to see how powerful a new bomb design is than by detonating one, after all. But they were also vulgar displays of prowess intended to intimidate the other side.[12]

Each blast lofted radioactive dust high into the atmosphere. The dust blew around the planet, and when it settled as fallout, it brought its radioactivity with it. There was so much fallout that background radiation levels rose slightly across the entire planet, and whisps of fission fragments – like strontium-90 – accumulated in people's bones and teeth. Background radiation and strontium-90 trended in lockstep with the number of bomb tests. (It's worth noting that the 0.11-millisievert increase in background radiation – tiny compared to the natural background global average of 2.4 millisieverts – was harmless.)

The fallout is still detectable today. If you scooped up a handful of soil from your garden and analysed it in a lab, you'd find amongst the insects about 10 trillionths of a gram of plutonium. There are atoms of bomb-derived strontium-90 in your bones and teeth right now, too, albeit in far smaller numbers than in the 1960s. Most of the lingering background radiation is from carbon-14, and it will wane as the isotope slowly decays over the coming millennia.[14]

The thickening fallout and growing number of weapons tests panicked politicians in the late 1950s. The USSR and the USA

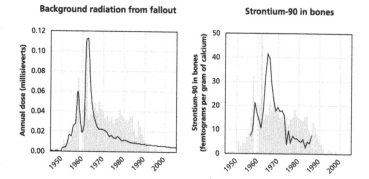

Global radiation dose from atom bomb fallout (left), and strontium-90 concentrations in the bones of 0–5-year-olds (right). I plotted these datasets against a backdrop of weapons tests. Tests after 1963 were conducted almost exclusively underground, and so their radiological effects were diminished.[13]

agreed to pause weapons testing at the end of 1958. But the hiatus was brief. In 1961 – the year the USSR detonated Tsar Bomba – testing recommenced with a renewed vigour, and in 1962, it peaked at an all-time high of 178 blasts. That's an atom bomb every other day for an entire year.[15]

And so it was fortunate that 1963 saw the penning of the *Treaty Banning Nuclear Weapon Tests in the Atmosphere, in Outer Space and Under Water*, more commonly called the Partial Test Ban Treaty. The USSR and the USA, along with the UK, committed immediately. As its name suggests, this treaty did *not* stop parties from testing bombs – just from testing them above ground. By the decade's end, more than 100 nations had committed; today there are 125. (France, China, and North Korea are the only atom-bomb nations not to have committed to the treaty.) Whilst testing continued with a steady cadence until the mid-1980s, fallout was trapped beneath the Earth where it can't harm people.[16]

As the USSR collapsed, weapons tests by the two world superpowers fizzled out, before stopping completely in 1992. In September 1996, the United Nations adopted the *Comprehensive Nuclear-Test-Ban Treaty*, which bans *all* tests unequivocally. It essentially makes it illegal to detonate an atom bomb, anywhere. To date, 178 nations have signed and ratified the treaty. Alas, it never came into force. Annex II identifies 44 nations – so-called 'Annex II States' – that possessed nuclear reactors at the time, all of which must sign and ratify the treaty for it to take effect. Of the 44 Annex II States, six have signed but not ratified it (China, Egypt, Iran, Israel, Russia, and the USA). Three haven't even signed it (India, North Korea, and Pakistan). And in 2023, Russia – which once fully embraced the pact – withdrew its ratification as its relations with the USA deteriorated in the wake of the Russo-Ukrainian War.[17]

Whilst the Comprehensive Test Ban awaits resolution, the overall picture since the frenzied 1960s is positive: atom bomb testing has declined enormously. Only one country has conducted tests this century – North Korea – and it receives international condemnation.

Taking stock

The world did not progress from atom bombs to nuclear power stations following World War II. It amplified both technologies simultaneously. By 1956 – the year the world got its first commercial nuclear power station – there were more than 4,000 atom bombs. By 1970, there were 77 power reactors and 38,000 bombs.[18]

Almost all the atom bombs that have ever existed belonged to either the USA or the USSR. As the Cold War tightened its icy grip on the world, the superpowers engaged in a reckless game of tit for tat by inflating their atomic arsenals. The world reached peak bomb in 1986 when the stockpile topped 65,000; 62 per cent of them belonged to the USSR, and 36 per cent to the USA. The remaining 2 per cent were divided amongst (in descending order) France, the UK, China, Israel, and South Africa.[19]

But after climbing for more than 40 years, the number of atom bombs declined rapidly through the late 1980s, 1990s, and 2000s, thanks largely to a series of agreements between the USA and the USSR/Russia. Whilst reductions slowed during the 2010s and there

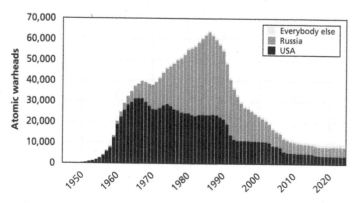

The precise numbers of atom bombs are state secrets, but these data represent best estimates. This graph doesn't include retired bombs awaiting dismantlement, of which there are (at the time of writing) about 2,500.[20]

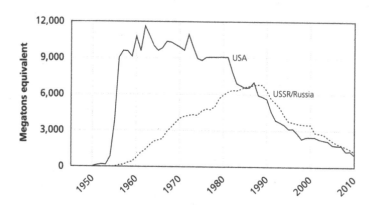

Estimates of the explosive energy of a first atomic strike. I omitted the other atom-bomb nations because their first strikes are too small to see on this scale.[22]

are still some 9,500 bombs in the global stockpile, they're down 85 per cent from the mid-1980s, which diminishes the ferocity of a first strike.[21]

I wouldn't call this 'great' news for humanity, exactly, because there are still enough atom bombs to end civilisation many times over. But the situation *is* improving overall.

In any case, I think it's far more likely that a nuclear war would be triggered accidentally than deliberately. A military aircraft carrying two hydrogen bombs dropped its payload when it disintegrated in mid-air over North Carolina in 1961; the trigger mechanisms of one of the bombs engaged, but detonation was averted by a last-ditch safety switch. In 1995, Russian officials briefly mistook a Norwegian rocket carrying instruments to study the aurora borealis for an American atomic missile; President Boris Yeltsin was about to launch a retaliatory strike before Russian surveillance realised it was a false alarm.[23] These broken arrows could conceivably cause twitchy fingers to hit the button, especially during politically tense times.

As with the frequency of weapons tests, the overall picture of atom bomb stockpiles is one of decline. By reducing the number of atom bombs in the world, we make false alarms less likely. The lesson from the late 1980s through the 2000s is clear: rapid nuclear disarmament is possible. It was unthinkable in the 1960s, but it happened. Reducing the stockpile further is achievable and ought to be a collective goal of our world leaders.

Non-proliferation

Today, there are nine atom-bomb nations: the USA, Russia, the UK, France, China, Israel, Pakistan, India, and North Korea. (Nobody knows for *sure* if Israel has atom bombs because it neither admits nor denies it has them, but expert consensus is that it almost certainly does.) South Africa acquired them in 1979 but disarmed completely in 1991.[24]

The USA's monopoly on the atom bomb broke when the USSR acquired their own in 1949. The UK acquired the bomb in 1952, France in 1960, China in 1964, and Israel (probably) in 1967. To halt the bomb's mushrooming, the United Nations opened the *Treaty on the Non-Proliferation of Nuclear Weapons* for signatures in 1968. With

191 parties today,* this treaty encompasses almost every nation, placing it in the ranks of the Constitution of the World Health Organization (193 Parties) and the founding Charter of the United Nations itself (also 193). The lifetime of the Treaty was extended indefinitely in 1995.[25]

The treaty rests on three pillars:

1. **Non-proliferation.** If you have atom bombs, don't give them to anybody else or help anybody else make them. If you *don't* have atom bombs, don't accept them from anybody else and don't make them yourself.
2. **Peaceful uses of nuclear technology.** The peaceful applications of nuclear technology – like nuclear power, nuclear medicine, and atomic gardening – should be available to all nations.
3. **Disarmament.** Nations with atom bombs should relinquish them for good.

It's the most widely adopted arms control treaty in the world, with only India, Israel, Pakistan, and South Sudan (and possibly North Korea) not party to it. India, Israel, and Pakistan would have to disarm to be allowed to accede, because the treaty restricts atom-bomb nation status to those that acquired atom bombs before January 1967. It's the same path South Africa took towards accession in 1991.[26]

Iran is party to the treaty but dances perilously close to flaunting it. In December 2023, it came to light that Iran had scaled up production of uranium enriched to 60 per cent. Since uranium enriched to 5 per cent is ample to kindle a light-water reactor, highly enriched uranium is of little use in nuclear power generation, unless you use it to kickstart a breeder reactor. The move was widely denounced on the international stage, with the British government describing it as an 'unabated escalation'.[27]

And therein lies a potential contradiction in the treaty: many of the

---

*Or 190, if you discount North Korea, which announced its withdrawal in 2003. Remaining parties cannot agree whether North Korea formally left or is technically still part of it.

nuclear power technologies can be applied directly to bomb-making. If a nation can enrich uranium, recycle plutonium, or breed fissile isotopes, it can go a long way towards making a bomb. Advanced nuclear power programmes make clandestine atom bomb production more likely . . . but we can't deny nations the right to develop these technologies because it violates the treaty's second pillar. In 2009, then Director-General of the International Atomic Energy Agency Mohamed ElBaradei described the conundrum as the 'Achilles' heel of non-proliferation'.[28]

The International Atomic Energy Agency was founded, in part, to resolve this catch. Since its inception in 1957, it has acted as the world's nuclear watchdog. An international team comprising hundreds of specially trained inspectors tours the world's nuclear facilities, checking that nobody is siphoning off fissile substances into illicit weapons programmes. The agency has been successful in its mission. It has sniffed out every clandestine nuclear programme in participating countries. (Of course, if a country had managed to covertly develop atom bombs, we wouldn't know about it, but it seems incredibly unlikely that a finished atomic arsenal would be kept under wraps; secret bombs aren't a deterrent.)

An entire branch of nuclear forensics supports the International Atomic Energy Agency in sleuthing the origins of spent fuel. If spent fuel is found somewhere it shouldn't be – outside of regulatory control, for instance – its isotopic fingerprint gives clues about exactly where it came from. Analysts make the measurements in a lab – there's one in Aldermaston, Berkshire – using instruments such as mass spectrometers, and the fingerprints are compared to isotopic libraries and computer simulations to elucidate what type of reactor it came from, how long it was in the reactor, and even how old it is. These sensitive pieces of information help triangulate potential motives and, when necessary, bring criminals to account.

Since 1993, the International Atomic Energy Agency has logged 4,075 cases of criminal or unauthorised activities involving nuclear or radioactive material on its Incident and Trafficking Database. Most incidents involved material that went missing, was stolen, or was disposed of inappropriately. Fewer than 10 per cent related to trafficking or malicious activity.[29]

Another resolution to the treaty's potential contradiction might be the establishment of international nuclear fuel centres: fuel fabrication and recycling facilities that are owned, operated, and shared by multiple nations. Nuclear reactor fuel can't be exploded in an atom bomb because it's not enriched enough, and so it practically eliminates the risk of bomb-grade material being diverted for nefarious purposes. Having multiple centres dotted around would insulate the world's supply of fuel from regional turbulence and the ebbs and flows of provincial politics. It would also save wannabe nuclear nations from establishing indigenous fuel cycles. All nations – developed and undeveloped, armed and unarmed – ought to have access to nuclear fuel. And their scientists and engineers ought to have the opportunity to work in these shared facilities. We should strive to make nuclear power available to everybody, whilst continuing to curtail the spread of atom bombs. Our chances of reaching net zero whilst working towards a bomb-free world hinge upon it.

And then there's the small matter of disarmament. Whilst the number of atom bombs has plummeted since the mid-1980s, only South Africa has disarmed completely. Atom-bomb nations aren't defying the *letter* of the treaty – it outlines no disarmament deadline – but they're defying its *spirit*. Possessing hundreds or thousands of atom bombs whilst actively preventing other nations from acquiring them exposes them to accusations of hypocrisy. It's incumbent on them to restart the trend of stockpile reduction in tandem.

In the face of these challenges, the Non-Proliferation Treaty has been successful. Eighty years since Gadget, 35 nations have developed nuclear reactors.* Despite there being many dozens easily capable of

---

* The 35 nations that have (or had) nuclear reactors are Argentina, Armenia, Belarus, Belgium, Brazil, Bulgaria, Canada, China, the Czech Republic, Finland, France, Germany, Hungary, India, Iran, Italy, Japan, Kazakhstan, South Korea, Lithuania, Mexico, Netherlands, Pakistan, Romania, Russia, Slovakia, Slovenia, South Africa, Spain, Sweden, Switzerland, Taiwan, the UAE, the UK, and the USA. As I write, three more nations – Bangladesh, Egypt, and Turkey – are building their first.

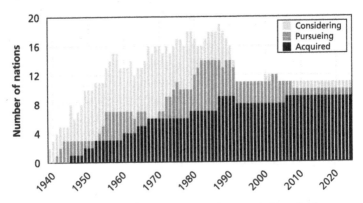

Nations that are considering, pursuing, or have acquired atom bombs. Iran is actively pursuing them; Syria is considering it.[30]

making them, only nine nations have atom bombs, and interest in acquiring them has generally waned. It's perfectly possible to have a world of one without the other.

Megatons to Megawatts

As the Cold War thawed in the early 1990s and the stockpile of atom bombs shrank, the world was presented with a new (but admittedly nicer) problem: what to do with all the nukes. This was especially pertinent in the disintegrating USSR, where bankruptcy and political turmoil meant a 500-tonne stockpile of highly-enriched uranium – enough to make some 20,000 atom bombs – wasn't secure. Many in the West feared it might be stored haphazardly in run-down bunkers under the care of corrupt and unreliable guards. The proliferation risks provoked intense anxiety. The last thing the world needed was for weapons-grade uranium to go missing, or worse, end up in the hands of terrorists or a rogue state.[31]

This is when Thomas Neff – a researcher at MIT – came up with a bright idea by going back to basics: 500 tonnes of weapons-grade uranium contains vast quantities of fissile uranium-235. That's the same isotope used to generate electricity in nuclear power stations. Why not put it to that exact use? In 1991, Neff wrote a *New York Times* op-ed that outlined what he called a 'grand uranium bargain': repurpose the USSR's highly enriched uranium stockpile by turning it into nuclear fuel and sell it to American power stations. It's an argument so obvious that, once realised, it writes itself.[32]

Scientists and politicians from both nations received the plans enthusiastically. Just 16 months later, the USA and (by that time) Russian governments signed the *Agreement Between the Government of the United States of America and the Government of the Russian Federation Concerning the Disposition of Highly Enriched Uranium Extracted from Nuclear Weapons*. Russia was to blend its highly enriched uranium with uranium-238, diluting its potency from weapons-grade (90 per cent uranium-235) to reactor-grade (4 to 5 per cent uranium-235), and the USA was to burn it in nuclear reactors. The deal became known as 'Megatons to Megawatts'. It was a win-win-win: cash-strapped Russia exchanged a liability for money, the USA got enormous amounts of nuclear electricity, *and* the world became safer.[33]

Between 1993 and 2013, the USA generated 8.5 *million* gigawatt-hours of electricity from the repurposed uranium. That's 10 per cent of its total demand. Not only was that electricity reliable, but it was also clean. If the USA had generated those 8.5 million gigawatt-hours of electricity from gas, it would have emitted an extra 4 billion tonnes of carbon dioxide; if it had generated it from coal, it would have emitted an extra 7 billion tonnes. That's comparable to the annual carbon dioxide emissions of Africa, Central America, the Middle East, Oceania, and South America combined (about 6 billion tonnes).[34]

Megatons to Megawatts concluded in 2013 after the entire stockpile had been repurposed. It's widely praised as one of the most successful non-proliferation endeavours ever undertaken. It's worth remembering that just 7 years prior to the programme's inception, the world was at peak bomb, with the USA and the USSR alone having 63,500 of them, mostly pointing at each other. But the uranium destined for detonation above American cities ended up powering them.

I'd love to see another programme like Megatons to Megawatts in my lifetime. We could reduce the number of atom bombs in the world *and* generate copious quantities of emissions-free electricity. As Morrissey sang over Johnny Marr's jangly guitar in 1986, the year of peak bomb, *if it's not love, then it's the bomb that will bring us together.*

Uninvention

Gadget exploded closer in time to the reign of Queen Victoria than to today. It was closer to the invention of the ballpoint pen and the gramophone record than it was to the iPhone. The atom bomb was ahead of its time; we were unprepared, in what Carl Sagan called our 'technological adolescence'. We can't uninvent it; the genie is out of the bottle, and since 1945, politicians, scientists, artists, and philosophers have been figuring out to do about it.

Developing the atom bomb in the first place was necessary. Glen Seaborg, who discovered plutonium, the chemical element that made Gadget and the bombing of Nagasaki possible, elaborated in 1990:

> There are young people today who feel that we shouldn't have developed the atomic bomb, that it was a mistake. I believe that this is because – through no fault of their own – they don't have this sense of history. They didn't live through this almost terrifying period when we thought we were losing the race with Adolf Hitler.[35]

In a race between the USA and Nazi Germany to develop the atom bomb, I'm glad the Americans won. 'We had understood full well what it would have meant if Adolf Hitler got the atomic bomb before the Allies did,' said Seaborg.[36] Whether it was right to drop two atom bombs on Japan is a different question.

A month before Gadget was detonated, seven high-profile scientists working on the Manhattan Project signed the then-secret but since declassified 'Franck Report'. Named after its lead author – Nobel Prize-winning physicist James Franck – the document warned against the bombing of Japan, advising the US government should frighten its enemy into surrender with a 'demonstration in an appropriately selected uninhabited area'.[37] Seaborg was amongst the signatories and later said:

> We were hoping that there could be a demonstration that Japanese representatives would attend and see the explosion. Anybody who saw the explosion of a nuclear weapon would realise what it would do to one's own people. And we thought then, those representatives would go back to the Emperor and the leaders of Japan and tell them 'we'd better surrender'.[38]

The 'Franck Report' also warned about other nations frantically acquiring the bomb, foreshadowing the arms race. The other argument – the one that prevailed – was that the atom bomb would end World War II swiftly, and more people would be killed in a drawn-out land invasion than by levelling two or three Japanese cities.

An atomic deterrence between atom-bomb nations has existed since the late-1940s. The fear of furious vengeance stopped any of them making a first strike. In 1985 – the year before peak bomb – US President Ronald Reagan and leader of the USSR Mikhail Gorbachev said 'a nuclear war cannot be won and must never be fought' in a joint statement. The threat of mutually assured destruction was enough to steady their hands. A joint statement in 2022 by the governments of China, France, Russia, the UK, and the USA echoed this sentiment: 'We affirm that a nuclear war cannot be won and must never be fought.'[39]

It's difficult to believe against the backdrop of the 24-hour news cycle of doom and gloom, but the world has enjoyed a period of unprecedented tranquillity since the end of World War II. War and deaths from armed conflicts are rarer now than they were in the nineteenth and twentieth centuries. Relationships between nation states are – on the whole – more peaceful than they've ever been. The spread of democracy and the growth of global trade no doubt played their parts, but it might also be the case that atom bomb deterrence, by frightening the world into peace, had a hand too.[40]

At the Tenth Review Conference of the Parties to the Treaty on Non-Proliferation of Nuclear Weapons in 2022, Mayor of Nagasaki

Tomihisa Taue urged atom bomb nations to fulfil their obligations to disarm in good faith. He said, 'Hiroshima will forever remain the first wartime atomic bombing site, but whether Nagasaki will remain the last wartime atomic bombing site depends on the future that we create.'[41]

Knowledge has no morals. It just *is*; it lacks agency. We humans, however, *do* have morals, and we do have agency. What to do with knowledge of the atom's intricate nucleus – whether to use it for the betterment of humankind or its torment – is down to us.

※ ※ ※

On 3 November 2006, a 46-year-old man named Edwin Carter arrived at a north London hospital with severe stomach pains, intense vomiting, and profuse diarrhoea. His pulse and blood pressure were normal, and he didn't have a fever. Doctors admitted him pending further investigations. They prescribed a course of strong antibiotics and hooked him up to an intravenous drip, but his symptoms persisted. Meanwhile, his blood cell and platelet counts crashed. Four days after being admitted, he told doctors he thought he'd been poisoned. He also revealed his true identity: his name wasn't Edwin Carter. It was Alexander Litvinenko, a former KGB agent who had defected from Russia's Security Service.

On 11 November – 8 days after arriving at the hospital – a raging fever enveloped Litvinenko. His blood cell and platelet counts continued to plummet. On the 13th, his hair fell out, and he started resembling someone undergoing chemotherapy or radiotherapy. But chemical tests came back completely normal, and a Geiger counter passed over his body returned a deafening absence of *clicks*. By the 16th, his white blood cells had all but gone. On the 17th, his bone marrow started failing. On the 18th, he turned yellow with jaundice. On the 19th, his heart rate became irregular. On the 20th, his kidneys started to fail. Still, the tests failed to identify a cause.

Doctors suspected radiation poisoning, despite the Geiger counter's silence. They sent samples of Litvinenko's blood and urine to Britain's Atomic Weapons Establishment for advanced nuclear testing. There, scientists smeared a drop of blood on a glass slide and half-covered it with a thin glass coverslip. They then exposed a sheet of radiography film to the slide. The film developed a foreboding opacity, proving Litvinenko's blood was radioactive. Only the film unobstructed by the coverslip developed opacity, proving the radiation was a type that could be blocked easily. It was alpha radiation. Blocked by a single layer of skin or a few centimetres of air, the Geiger counter had missed it.[42]

Next, British scientists set about identifying the isotope responsible. They placed a sample of Litvinenko's urine in a gamma-ray spectrometer and saw the tiny hints of a peak, with an energy of 803 kiloelectron-volts. One of them recognised the murder weapon immediately by its signature gamma-ray signal: polonium-210, a rare and intensely potent alpha emitter. Tiny quantities of polonium-210 carry enormous amounts of radioactivity. Six picograms – that's 0.000000000006 grams, six-millionths of a millionth – spits out 1,000 high-energy alpha particles every second. Litvinenko had ingested over 4 million times that amount.[43]

Litvinenko was suffering from acute radiation syndrome. A few hours after his poison was identified, he fell unconscious and died of heart failure. He never stood a chance.

The polonium trail

Polonium-210 is a clever method of assassination. It has no odour, no taste, and – with a millionth-of-a-gram lethal dose – no colour. The delayed death lets the assassin slip away and evade immediate capture. Despite its immense alpha radioactivity, it can pass quietly through airport radiation scanners because it emits feeble gamma rays and no beta radiation. It's also totally barbaric; an agonising death awaits, designed to incite terror. The vivid photograph of Litvinenko lying in his hospital deathbed – gaunt, jaundiced, and baled with alopecia – was printed widely across the western media following his death. Who did it? And how?[44]

Police investigators and nuclear scientists joined forces to find

Litvinenko's assassins. Thus ensued no ordinary forensic investigation, but a *nuclear forensics* investigation, where high-tech analytical instruments – gamma-ray spectrometers, mass spectrometers, and the like – were deployed alongside police interviews and fingerprint dusters.

Forensic investigators retraced Litvinenko's steps. On the day he fell ill, he'd met two former KBG agents named Andrey Lugovoy and Dmitry Kovtun in a central London hotel bar, where Litvinenko had ordered a green tea with lemon and honey. Investigators combed the bar with Geiger counters. Even weeks after the meeting took place, frenzied *clicks* revealed splotches of residual polonium-210 *all over the bar*. And there was lots of it. They'd found the scene of the crime.[45]

There was no smoking gun, but there was a porcelain teapot. It was riddled with polonium-210. The inside of the spout was so radioactive the Geiger counter's *clicks* revved into overdrive. Lugovoy and Kovtun must have spiked Litvinenko's drink.

Investigators followed 'the polonium trail' the assassins left in their wake in the fortnight leading up to Litvinenko's murder. It was in the hotel rooms Lugovoy and Kovtun slept in. It was in the offices they'd used to hold meetings, a restaurant they'd eaten in, and a nightclub they'd visited. On 3 November, the day Litvinenko arrived at the hospital, Lugovoy and Kovtun flew back to Russia aboard British Airways flight BA874. Row 16 of the cabin, where Lugovoy and Kovtun had sat, was contaminated with polonium-210, too.[46]

Nuclear scientists also analysed the amount of lead-206 – the non-radioactive progeny of polonium-210 – along strands of Litvinenko's hair using a mass spectrometer. The more lead-206 in the hair, the more parental polonium-210 there must have been at the time that section of hair grew; each hair is like a tape recorder that preserves information about the amount of polonium-210 coursing through the body over time. The region of hair closest to the root revealed the highest lead-206 concentration, corresponding to the fatal poisoning. But further along the strand – some 5 millimetres up from the root – there was another lead-206 spike. It looked like a previous attempt to poison Litvinenko. By estimating the speed at which hair grows, analysts approximated the date of the possible assassination as sometime between the 14 and 18 October. The date range lines up with a meeting in London between Litvinenko, Lugovoy, and Kovtun on the 16 of October. Litvinenko had also fallen ill that evening. The conclusion was inescapable: *there was an earlier, botched assassination attempt.*[47]

Naturally, the question arose: where did Lugovoy and Kovtun get the polonium-210? It's not the sort of poison a petty criminal can cook in his illicit laboratory or purchase on the black market. It's notoriously difficult to make and (evidently) dangerous to handle, and all aspects of polonium-210 production are strictly regulated and controlled by specialist government facilities. Every nuclear reactor in the world makes spent fuel, which makes deciphering its exact origin formidable. Single isotopes, on the other hand, come from just a few specially equipped labs, especially the rare ones such as polonium-210. The only way to alchemise it in copious quantities is by irradiating bismuth-209 with neutrons in a nuclear reactor, where it becomes bismuth-210 before beta decaying swiftly to the final product.

Norman Dombey – Professor Emeritus of Theoretical Physics at the University of Sussex – revealed a trail that stretched thousands of miles eastwards. It ended at a nuclear reactor in Mayak, Russia, where most of the world's polonium-210 is made. 'I know of no other large-scale production facility of Po-210,' he would say in the public enquiry. Although Dombey couldn't prove it definitively, he concluded it was 'probable' the polonium-210 was made in that facility and, if that were true, 'the Russian state or its agents were responsible for the poisoning.'[48] Individual vendettas are no match for collective nuclear science.

Pulling together more than 600 pieces of evidence, including more than 30 pieces of *expert scientific evidence*, the public enquiry into the assassination was published in 2016. It implicated Lugovoy and Kovtun. It also concluded the 'operation to kill Mr Litvinenko was probably approved by Mr Patrushev [head of Russia's security agency, successor to the KGB] and also by President Putin.' The European Court of Human Rights came to a similar conclusion in

2021 when it said there's 'a strong *prima facie* case that . . . Mr Lugovoy and Mr Kovtun had been acting as agents of the Russian State.' Litvinenko was a defiant and vocal critic of the Kremlin. Lugovoy and Kovtun deny involvement.[49]

A needle in the Outback

Nuclear forensics is also deployed in cases of hide-and-seek. Radioactive sources – such as those used in smoke alarms and hospitals – sometimes go missing. Since it started tracking them in 1993, the International Atomic Energy Agency has recorded more than 4,000. Just 8 per cent are stolen by illicit traffickers or malicious actors. Most sources are simply misplaced by accident.[50]

One of the most famous incidents of inadvertent losses culminated in a week-long episode of lost-and-found that gripped the world's media. In January 2023, a lorry disembarked from Rio Tinto's Gudai-Darri Mine in Western Australia. Bound for Perth, the gruelling 870-mile journey ran through vast desolate stretches of the outback. The lorry's cargo wasn't your usual run-of-the-mine equipment: it was 6 milligrams of radioactive caesium-137, the mass of about three sand grains. Six milligrams of caesium-137 packs a punch. With a half-life of 30 years, it's 125 *million* times more radioactive than an equivalent mass of natural uranium. And unlike alpha-emitting uranium – which is only radiotoxic if inhaled, injected, or ingested – it's potent at a distance; caesium-137 emits long-range beta radiation *and* gamma rays.[51]

Mining companies use caesium-137 as a density gauge by shining its intense gamma rays through ores and minerals. Rio Tinto's caesium-137 speck was sealed inside a silver capsule measuring 6 millimetres wide and 8 millimetres long, about the size of a peanut. The courier came to unpack the radiocargo 13 days after it left Gudai-Darri Mine . . . but the gauge was broken, and one of the bolts fixing it in place was gone. The caesium-137 was nowhere to be found. The capsule had wriggled itself loose of its container and slipped through a crack in the truck somewhere in the outback.[52]

Within 24 hours, Rio Tinto had alerted the authorities. The next day, an urgent public health warning was issued. The message was simple: if you find it, *stay away!* Authorities were terrified that somebody might discover it and stow it curiously away for safekeeping, exposing themselves to an insidious dose of radiation, skin burn, and radiation sickness. It was small enough to get stuck in the treads of a car tyre, too, raising the horrifying possibility that it might be carried into a densely populated city or somebody's garage. Radiation is harmless to almost everybody in any situation they ordinarily find themselves; getting close to six thousandths of a gram of caesium-137 without adequate shielding is a rare exception.[53]

The seemingly impossible search for the capsule began immediately. It was like hunting for a garden pea on the road between John o' Groats and Land's End. But scientists at the Australian Nuclear Science and Technology Organisation had a cunning plan. They modified a gamma-ray detector, about the size of a vacuum cleaner, and stowed it in the back of a car. Then, they retraced the journey, combing the lonely stretch of road for gamma-ray spikes. Day one of the search: nothing. Day two: nothing. Days three, four, and five: still nothing.

But on the sixth day, the alarm rang from the back of the car. 'It was very exciting . . . we knew that we had found it,' a physicist involved in the mission later said. The detector had registered a spike in 662 kiloelectron-volt gamma rays, the signature product of caesium-137 decay. Sure enough, they had found the tiny capsule with nothing but a car and a gamma-ray detector. 'We were absolutely stoked.'[54]

A carbon blast from the atomic past

The gases at the top of Earth's atmosphere face the full fury of cosmic rays from outer space. Collisions between the extraterrestrial incomers and indigenous air atoms generate an assortment of strange isotopes, such as radioactive carbon-14, which beta decays with a 5,700-year half-life. Its steady production and continuous decay are poised naturally in a state of balance; at any one time, roughly one in every trillion carbon atoms in Earth's atmosphere is carbon-14. Whilst it varies slightly over thousand-year timescales, the prevalence of carbon-14 remains more or less constant.

Constant, that was, until the 1950s. Like artificial cosmic rays, stray neutrons from atom bomb tests forged copious quantities of carbon-14. Mushroom clouds lofted it skyward, and in a matter of years, the amount of it in the atmosphere doubled. Its prevalence increased far beyond the one-in-a-trillion produced naturally by cosmic rays. This so-called *bomb pulse* climbed rapidly until the Partial Test Ban Treaty put an end to open-air bomb tests. After peaking in 1965, atmospheric carbon-14 dwindled as it was soaked up by the oceans and absorbed by vegetation. (Radioactive decay has contributed to its decline, but barely; it's got such a long half-life that it takes 83 years to wane by just 1 per cent.) The shape of the bomb pulse mirrors the first decade of the atomic arms race.[55]

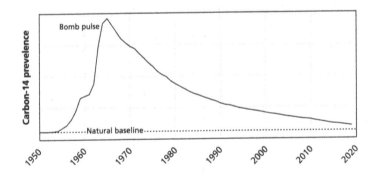

The bomb pulse left its trace in living organisms, too. Plants inherit the atmosphere's carbon-14 signature as they grow, and when animals eat those plants — and when other animals eat those animals — the carbon isotopic signature propagates through the food chain. There's a slight lag because it takes a while for the atoms in your body to be replaced by new ones, but the prevalence of carbon-14 in the air you're breathing right now closely reflects the prevalence of carbon-14 in *you*.

When a plant stops photosynthesising or an animal stops eating – i.e. when it dies – its carbon isotopic signature freezes; the carbon in it no longer changes in lockstep with the atmosphere. By measuring the prevalence of carbon-14 in a deceased organism with a mass spectrometer, and seeing what year it corresponds to on the bomb pulse, *the year of death can be determined*. Carbon isotopes persist decades after soft tissues have decayed. This dating technique is called *bomb pulse dating*, and it's sometimes used in forensic investigations to deduce the time of death.

Bomb pulse dating only works on the remains of those who died after the early 1950s. Before then, there was no bomb pulse to pinpoint their time of death. But carbon isotopes can still at least help determine whether somebody died before or after the pulse. In 2003, a corpse was chanced upon by builders in Panama as they excavated land close to an enclave of former dictator Manuel Noriega. Authorities suspected the remains belonged to a victim of foul play sometime during the political unrest of the 1980s, but traditional forensics failed to find a date of death. Ten years later, researchers in North Carolina measured the carbon isotope signature of the remains and found no traces of the bomb pulse. The person, therefore, died long before the 1980s and had nothing to do with Noriega or his dictatorship.[56]

Some parts of the body do not turn over their carbon. Tooth enamel, for instance, doesn't regenerate after it forms in childhood. Thus, the carbon-14 prevalence in a given tooth will always reflect the atmospheric carbon-14 levels when it formed. By measuring the carbon-14 in a tooth, and by knowing the individual's age when the tooth developed – which depends on whether it's a canine, an incisor, or a molar – *we can easily calculate the year the individual was born* by seeing where it lies on the bomb pulse. This method of dating is accurate to within 19 months. By contrast, determining the age of an adult skeleton from the bones alone has a give-or-take of 5 to 10 years. Like an isotopic birth certificate, your precise age is permanently inscribed in your teeth. (Unless you were born pre-bomb pulse, in which case you could have been born in 1940 or 1740.)

There's a slight complication in calculating somebody's birth year if it's unclear whether they were born before or after the peak of the bomb pulse; their carbon-14 signature could correspond to a year on the *rising* face of the slope or the *falling* face of the slope. In such cases, two teeth must be measured, and so long as they formed at different times during childhood, their relative positions on the pulse – and thus whether they fall before or after its peak – can be deduced.[57]

Measuring a person's age this way is incredibly useful for identifying human remains, especially when they're in a bad way. It helped identify victims of the tsunami in the Indian Ocean on Boxing Day 2004 whose bodies were too decomposed to tell teenagers apart from elderly people. And in 2013, researchers used it to help identify the body of someone who had been shot in the face and dumped in an artificial lake in northern Italy.[58]

Bomb pulse dating shouldn't be confused with *radiocarbon dating*, also known as *carbon dating*. The former applies to events that occurred after the early 1950s; the latter relies on measuring radioactive decay that's taken place naturally over the hundreds or thousands of years since an organism died, and can only be used for *archaeological* remains, not recent ones.

Aside from helping solve murder mysteries and identify bodies, bomb pulse dating helps in the fight against wine forgeries.[59] Grapes inherit the atmosphere's carbon-14 signature when they grow on the vine and pass it on to wine when they're crushed and fermented. There's no hope in trying to disguise a young Cabernet Sauvignon as a well-aged bottle, because carbon isotopes will reveal a wine's vintage down to the year. A counterfeit might get past a connoisseur, but there's no fooling a mass spectrometer. *In vino veritas*, as they say, but sometimes you need a nuclear chemist to uncover it.

※ ※ ※

The collapse of the Soviet Union presented the world with a new threat: nuclear materials, smuggled from former Soviet states, might make their way into the hands of criminals. Law enforcement authorities and police investigators suddenly developed a keen interest in nuclear analysis. If they found nuclear material in the wrong hands, they wanted to know exactly what it was and where it came from. Thus, a forum for informal collaboration called the Nuclear Forensics International Technical Working Group – open to all states interested in advancing nuclear forensics – was established in 1995.

The working group hones nuclear forensic methods and strengthens the ties between scientists and law enforcement. It holds an annual meeting, shares best practices amongst members, and stages atomic crimes so labs can practise their sleuthing skills. And, as its name suggests, it transcends national borders. Since its inception, hundreds of scientists spanning almost 60 nations have been involved. It publishes its activities in the open scientific literature and a publicly available quarterly newsletter (in, incidentally, both English and Russian).[60]

The working group – which involves everybody from nuclear physicists to police inspectors and encompasses research bodies worldwide – is emblematic of nuclear forensics in general: it's inherently transparent, interdisciplinary, and collaborative. It rallies people around a common goal, bringing them together as it makes the world safer.

Chapter 13. The Final Frontier

There's a starman waiting in the sky,
He's told us not to blow it,
'Cause he knows it's all worthwhile.

– 'Starman', David Bowie, 1972

For most of human history, mariners sailed the oceans of Earth by starlight. Stars were faithful beacons by which to navigate through treacherous waters. Guided by the night sky, our forebears voyaged over the horizon to new shores, and in time, we reached every landmass and connected the entire world.

In 1957, with no new horizons to sail over, we set our sights on an extraterrestrial frontier. We sent our first robotic emissary, the USSR's *Sputnik 1*, into orbit. In 1961, our species became spacefaring when the first cosmonaut – Yuri Gagarin – circumnavigated Earth; by decade's end, four astronauts – Neil Armstrong and Buzz Aldrin of Apollo 11, and Pete Conrad and Alan Bean of Apollo 12 – had left footprints on the Moon. Cosmonauts are named after the ancient Greek κόσμος ('cosmos') for *universe* and ναύτης ('nautes') for *sailor*; they are *universe sailors*. Astronauts are named similarly after ἄστρον ('astron') for *star*; they are *star sailors*. Our spacefaring ambassadors – robotic and human alike – voyage across the sea of space in vessels powered not by paddle or wind, but by rocket fuel.

In the late 1970s, the outermost planets – Jupiter and Saturn the gas giants, and Uranus and Neptune the ice giants – aligned, tracing an arc that swept from the Sun towards the edge of the solar system. Such geometric arrangements only happen every 175 years. This one presented us with a rare opportunity: we could gravitationally skip from one planet to the next, performing a grand tour of the outer solar system with minimum fuel and in minimum time; an encounter with one planet would bend a spacecraft's trajectory and give it a boost towards the next. A direct journey from Earth to Neptune normally takes 30 years, but the planetary alignment would permit it in just 12. The appropriately named *Voyager 1* spacecraft was thereby dispatched from the launchpad in Cape Canaveral, Florida in September 1977. First stop: the Jovian system.[1]

Voyager 1 arrived at Jupiter in March 1979 after an 18-month crossing. Jupiter's striped atmosphere and Great Red Spot — a vortex large enough to swallow Earth several times whole — swirled as the spacecraft flew by. Boosted by Jupiter's gravity, *Voyager 1* arrived at its next destination — the Saturnian system — 20 months later to an exquisite system of rings and medley of moons. Eighteen hours before its nearest approach to Saturn, *Voyager 1* flew close to its largest moon, Titan, shrouded in a smoggy atmosphere of nitrogen and hydrocarbons. Sailing 4,000 miles above Titan's orange cloud tops — the same distance as from London to Chicago — *Voyager 1* peered down at a potentially habitable world. But the close encounter came at a high price. Titan's gravity deflected *Voyager 1* off course. *Voyager 1* picked up enough speed to escape the solar system and sped northwards into the interstellar abyss.[2]

But as *Voyager 1*'s name suggests, there was a second spacecraft just behind. And *Voyager 2* did make it to Uranus in January 1986 and then Neptune in August 1989. It remains the only spacecraft ever to visit the ice giants. Together, the *Voyager* twins completed a full reconnaissance of the outer solar system and rewrote the astronomy textbooks.

The *Voyager* spacecraft were designed meticulously. Spectrometers, cameras, and cosmic ray detectors sat atop masts protruding from their keels, and 12-foot communication dishes relayed spools of data back to mission control in Pasadena, California. But a casual glance reveals a lack of something that is almost ubiquitous on spacecraft: solar panels. The *Voyager* twins were not solar-powered because the Sun is too feeble in the outer solar system. Where, then, did they get their power?

256 *Going Nuclear*

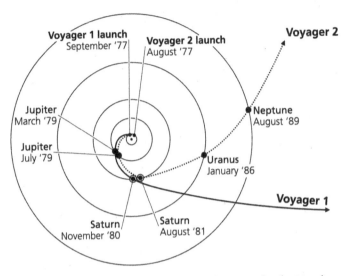

A bird's-eye-view of the solar system showing the Voyagers' paths. Note that Voyager 2 launched 1 month before Voyager 1 but was soon overtaken.

Plutonium power

Voyager was battery-powered. It wasn't powered by a chemical battery like the ones that might electrify your phone for a day or push your car along for a few hundred miles. A chemical battery couldn't power a 12-year trip to Neptune. No, *Voyager* had a *nuclear battery*.

The nuclear battery aboard *Voyager* ran on one of plutonium's rarest isotopes: plutonium-238, which spits out energetic alpha particles. A single gram of plutonium-238 emits about half a watt of thermal power through spontaneous alpha decay. Its radioactivity makes it self-heating. When pressed into a ceramic in oxide form (PuO_2, plutonium dioxide, to be precise), a grape-sized sphere has the same heat output as a tealight candle and glows cherry-red all by itself.[3]

Voyager had 13.5 kilograms of plutonium-238 on board, enough to fill about three coffee cups. It collectively churned out 7.2 kilowatts of thermal power, about thrice the power of a typical toaster. But heat alone doesn't power spacecraft; they need *electricity*. To that end, *Voyager*'s nuclear battery dutifully converted the plutonium decay heat into electric power using a phenomenon called the *Seebeck effect*: a temperature difference – between the inside, heated by the natural decay of plutonium-238, and the outside, cooled by the frigidity of outer space – generated a voltage that electrified the spacecraft's circuitry.* With no moving parts, *Voyager*'s nuclear battery easily withstood the trauma of launch and generated electricity reliably once in space, completely independently of the Sun.[4]

Voyager's nuclear battery generated 470 watts of electrical power at launch, enough to power 50 or so lightbulbs. Whereas a chemical battery depletes in a matter of hours, lumps of plutonium-238 cool down in tandem with its 88-year half-life, and so stay hot for centuries. Nuclear batteries are unsurpassed in their longevity. By the time *Voyager 1* reached Jupiter 18 months after launch, its battery was still 98.8 per cent full; by the time it reached Saturn, it had waned to 97.5 per cent. And by the time *Voyager 2* reached Neptune, its final destination, 12 years and 5 days after launch, its battery was still at 91 per cent power. Can you imagine if your phone's battery lasted that long?[5]

Of course, you wouldn't power everyday tech with plutonium-238, because, gram for gram, it's 25 million times more radioactive than natural uranium. It's also incredibly rare. Plutonium-238 doesn't exist naturally. Nor does it exist in spent nuclear fuel in any meaningful quantities. The only way we can get our hands on it in the kilogram-quantities needed for space batteries is by intentionally alchemising it in a multi-stage process. First, nuclear chemists shower

---

* In the 1970s, American physicians implanted plutonium-238-powered pacemakers into hundreds of cardiac patients. The advanced and easily rechargeable batteries we enjoy today didn't exist back then; miniature nuclear batteries solved the problem because they don't need to be recharged for decades. The typical radiation dose to a patient was around 5 millisieverts annually, twice the natural background dose, and less than what you'd get naturally in Finland or Cornwall. They're not used today, but when an individual dies who has one, the pacemaker is sent to Los Alamos where the plutonium is recovered.

uranium-235 seeds with neutrons inside specially designed nuclear reactors. Whilst uranium-235 atoms normally fission under a neutron deluge, sometimes they *absorb* the neutrons, swelling to become the next-heaviest isotope, uranium-236. And then the uranium-236 absorbs a neutron in turn, swelling to uranium-237, which beta decays swiftly to neptunium-237.

Chemists then chemically purify the neptunium-237 from a radioactive medley of uranium and fission fragments. It's laborious work but yields a feedstock of neptunium-237 that can be put back into a reactor and showered with neutrons once more. The neptunium-237 precursor swells to neptunium-238, which beta decays swiftly to space battery fuel.[6]

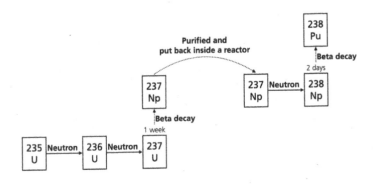

Much of the work is performed by chemists using robot hands from behind the safety of thick concrete. In the final stage, they purify the precious plutonium-238 from the neptunium, combine it with oxygen to form plutonium dioxide, and then sinter it into a tough ceramic pellet. Then, it's destined for space.

New horizons

First launched into orbit around Earth aboard military navigation satellites in the early 1960s, plutonium-238 has since taken us to every

Apollo 12 astronaut Alan Bean lifts a plutonium-238 fuel element from the Apollo Lunar Module. The finned object by his knees is the battery's casing. Image courtesy of NASA.

planet apart from Mercury. Unshackled from solar dependency, plutonium-238 lets us go anywhere.

In November 1969, the Apollo 12 astronauts touched down in Oceanus Procellarum, 'the Ocean of Storms'. The moonwalkers took with them a 74-watt nuclear battery, which they used to power their science experiments. Nuclear space batteries flew aboard every Apollo mission that followed and, except Apollo 13, which never *quite* made it to the lunar surface, remain on the Moon to this day. Apollo 13's nuclear space battery still rests at the bottom of the Pacific Ocean.

On 20 July 1976 – 7 years to the day since Neil Armstrong took his famous giant leap for mankind – we stepped onto the surface of another extraterrestrial landscape. Or rather, the *Viking 1* lander pressed its mechanical feet into the Martian surface on our behalf. Shortly after a soft touchdown, *Viking 1* snapped the first photograph ever taken on the surface of the Red Planet and gave the first Martian weather forecast: a low of −88 °C in the early hours but climbing to −31 °C by mid-afternoon, with moderate south-westerlies prevailing throughout the day. *Viking 1* was joined by its twin, *Viking 2*, a fortnight later. Both landers were plutonium-powered, as coverings of Martian dust render solar panels useless.[7]

Since *Viking*, we've sent two more plutonium-powered landers to Mars, first *Curiosity*, which landed in 2012, and then *Perseverance*, which landed in 2021. Both landers are minibus-sized laboratories equipped with spectrometers, cameras, and – crucially – wheels. As I write, *Curiosity* has roved almost 20 miles from its landing site in Gale Crater.

Further from Earth, *Galileo* flew 10,000 miles above Venus in February 1990 for a gravity boost on its way to Jupiter. During its 34 orbits around Jupiter, *Galileo* found evidence of oceans of liquid water beneath the icy crusts of Jupiter's moons and transformed our understanding of what lay beneath Jupiter's upper atmosphere. Almost 14 years after launch, scientists ended the mission by steering the spacecraft into Jupiter where it was crushed to oblivion by the overwhelming pressure.

Further still, *Cassini* slipped into orbit in 2004 around the jewel of the solar system: Saturn. During its 13 years in orbit, *Cassini* took thousands of photographs and made millions of measurements. It deciphered the fine structures in Saturn's rings and discovered ice volcanoes emanating from Saturn's moon Enceladus. It also picked up where *Voyager* left off in the exploration of Titan; *Cassini* revealed a world of organic chemistry where hydrocarbon rivers, fed by deluges of hydrocarbon rain, flow into vast hydrocarbon oceans. At the end of its life, *Cassini* was steered into its host planet, where it met a fate similar to *Galileo*'s. Spacecraft are often crashed to preserve the purity of worlds that might harbour life; the last thing we want to do is contaminate extraterrestrial biospheres with the earthly microbes on our spacecraft.

Chapter 13: The Final Frontier 261

Plumes of ice – photography by Cassini – gush from cracks in Enceladus' icy crust, hinting that an ocean of liquid water lies beneath.
Image courtesy of NASA/JPL-Caltech/SSI.

In 2015, *New Horizons* flew by Pluto. Out there, the Sun is almost 1,000 times dimmer than it is here on Earth. *New Horizon's* sensitive cameras captured Pluto's frozen canyons and precipitous highlands, showing that the dwarf planets are far from desolate backwaters – they're every bit as geologically and chemically sublime as the inner planets. And what fuel could get such a spacecraft out to the edge of the solar system and transmit photographs back to Earth but plutonium? How fitting that the chemical element that made *New Horizons* possible was named after the planet that it explored.

The dozens of nuclear batteries launched from Earth since the 1960s have enabled the full exploration of the solar system. And there's more to come. NASA is planning to send another spacecraft to Titan in 2028 named *Dragonfly*. It will soft-land on the hydrocarbon moon with the explicit aim of characterising its suitability for harbouring life. It will be equipped with a mass spectrometer, a gamma-ray spectrometer, and helicopter blades. The so-called 'rotorcraft' will fly hundreds of miles through Titan's smoggy atmosphere looking for the chemical signs of life. Powered by plutonium-238, it might help answer one of the oldest questions: *are we alone in the cosmos?*

Beyond *Dragonfly*, the possibilities are endless. Why not use the decay heat of plutonium-238 to melt a lander through the frozen shells of icy moons to the oceans of liquid water beneath? Or why not use a nuclear battery to power a blimp-like spacecraft that floats around in the atmosphere of an outer planet? Or perhaps we could even design plutonium-powered ships and submarines to sail and swim through the hydrocarbon oceans of Titan. With plutonium-238, not even the sky's the limit.

The fuel of the future

But there's a huge problem: the world's supply of plutonium-238 is running perilously thin. Almost all of it was alchemised decades ago in the USA as a by-product of atom bomb production. At the end of the Cold War, production ceased, and much of the remaining stockpile has since been blasted into outer space, or has decayed naturally with the passage of time. Russia – the other major producer of plutonium-238 – has almost run out, too. Without plutonium-238, we have no fuel for space batteries; without space batteries, we're stuck with solar-powered missions. The surface of Mars and regions beyond Jupiter are off-limits. What to do?[8]

A good place to start is with what we've got. The world has thousands of tonnes of plutonium bound up in spent nuclear fuel from power stations. Close inspection of this so-called 'reactor-grade plutonium' using a mass spectrometer reveals it's made principally from four isotopes: plutonium-239 (the lightest), plutonium-240, plutonium-241, and plutonium-242 (the heaviest). Alas, none of these isotopes make good fuel for space batteries. Three of them (plutonium-239, plutonium-240, and plutonium-242) decay too slowly to produce sufficient heat to drive the Seebeck effect. The fourth, plutonium-241, decays with a relatively short half-life of just 14 years, but decays by emitting *beta* radiation, dissipating its energy throughout its surroundings rather than in a concentrated lump.

However, the radioactive decay product of plutonium-241 – americium-241 – *is* good for space batteries. Whilst gram for gram it's five times less radioactive than the traditional plutonium-238 space battery fuel, it's still potent enough to self-heat. Its alpha radioactivity makes it plenty hot enough to generate electricity.

The problem with americium-241, as so often with esoteric isotopes, is acquiring it in large enough quantities. You only need a few hundred nanograms of americium-241 to make a smoke alarm, but you need tens of thousands of billions of nanograms – tens of *kilograms* – to fabricate a space battery. To find that much americium-241, we need to look in old stockpiles of reactor-grade plutonium. 'Old' is of paramount importance; too young, and not enough time will have passed for the plutonium-241 to decay into its americium-241 progeny. Like a fine vintage, reactor-grade plutonium gets better with age.

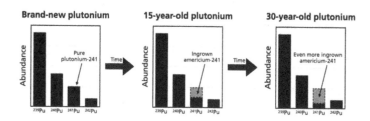

A mass spectrum of reactor-grade plutonium. Only plutonium-241 has a short enough half-life to decay in human timescales. It's replaced by its americium-241 progeny as it ages. (The isotopic fingerprint of reactor-grade plutonium varies between nuclear reactors; my sketch represents a typical light-water reactor.)[9]

Where on Earth could we find an old stockpile that's been idle for decades? Perhaps a stockpile that was separated from spent nuclear fuel but never recycled into new nuclear power? One place in particular springs to mind: *Sellafield*. Whilst the British government has been agonising over what to do with the UK's 141-tonne 'zero value asset', the plutonium has been maturing. A back-of-the-envelope calculation reveals there's now somewhere between 5 and 15 tonnes of americium-241 at Sellafield, with more and more each day. That's far more than all the plutonium-238 that's ever flown in outer space combined. Even the lower end of that estimate is a virtually inexhaustible supply of fuel for space batteries, enough to power at least 80 Voyager missions. It's far easier (and cheaper) to make than plutonium-238 and, short of it being recycled into MOX fuel, puts 'waste' plutonium to good use.[10]

With that in mind, my colleagues at the United Kingdom National Nuclear Laboratory are chemically mining a small portion of the stockpile for the precious americium-241. They're purifying it from aged plutonium with the explicit aim of bundling it onto a spacecraft and sending it into outer space. First stop? Probably Mars. In 2030, the European Space Agency will land its *Rosalind Franklin* rover on the surface of the Red Planet. In addition to its array of scientific instruments, *Rosalind Franklin* will be equipped with a 2-metre-long drill, which it will use to burrow beneath the red sands in search of past Martian life. And whilst the rover will be solar-powered, it will carry with it a small lump of americium-241 from Sellafield. It will keep some of its components warm. *Rosalind Franklin* will be Europe's first Mars rover and will carry with it the first americium-241 ever flown aboard a spacecraft.[11]

Of all the uses for nuclear 'waste', recycling it into the next generation of nuclear space batteries has got to be one of the most inventive.

Moon bases and beyond

Mars rovers such as *Rosalind Franklin* will search for the signs of ancient extraterrestrial life. But what about *future* life – *human* life – in our solar system?

To support the necessities of life on the Moon – fresh air, drinkable water, lights, heating and cooling, pressurisation, communication systems, transportation – we need more than the few hundred watts of electrical power generated by a space battery. We need tens of thousands of watts, or perhaps even *millions*. And, crucially, those watts must arrive reliably. A blackout on the Moon would be a death sentence for Moon-dwellers. Solar panels or chemical batteries won't suffice because the bitter lunar nighttime lasts for a fortnight. To keep people alive and well on the Moon for weeks and months, we have one choice: *bona fide* nuclear reactors.

Close to the Moon's south pole lies a bowl-shaped crater named after the great Antarctic explorer, Ernest Shackleton. Craters like Shackleton sit at such low lunar latitudes that the Sun never rises high enough to crest their rims. Their bottoms are permanently shadowed. The darkness holds their frigid floors at less than −160 °C, and over billions of years of solar system history, they've accumulated vast amounts of ice, probably from passing comets that pitter-pattered down shards as they streamed by. One of Shackleton's neighbours, Cabeus, contains an estimated 500 million tonnes of water, about as much as there is in Sydney Harbour. The profusion of water in the Moon's permanently shadowed craters makes them excellent candidates for the first human outposts on the pockmarked lunar surface.[12]

I envisage a future where humanity will thrive on the Moon by deploying micro-nuclear reactors along the rims of craters like Shackleton and Cabeus. Moonwalkers will use the power from those reactors to mine the frozen water on the crater floors and haul it up onto the crater rims, where they'll melt it using the nuclear reactor's warmth. They'll drink the water and use it to irrigate the gardens inside their lunar habitats. Using nuclear electricity, they'll pry the water into its elemental constituents: oxygen, which they'll breathe; and hydrogen, which they'll use to concoct rocket fuel for their return trips home ... or their onward journey into the rest of the solar system. Nuclear space batteries will play their part, too; perhaps they'll power instruments and provide underfloor heating. And once we've got the hang of living on the Moon, we'll venture further from the Sun and set up habitats on the surface of Mars. The journey will first be taken by specially trained astronauts, just as Antarctica was once frequented only by daring explorers. But in time, Mars will welcome tourists and resident scientists, just as Antarctica does today. Eventually, there will be hotels, casinos, and day spas. One day, there will be permanent Martian outposts. The first Martian cities will grow thereafter. Nuclear power will enable humanity to become an interplanetary species.

Such visions sound like the stuff of science fiction, but nuclear scientists were thinking about nuclear-powered Moon bases before Neil Armstrong and Buzz Aldrin ever set foot there. During a speech at an event celebrating what would have been Marie Curie's 100th birthday in Poland, 1967, Glen Seaborg said:

> In some ways, I believe establishing and maintaining a manned colony on the Moon will provide the most interesting challenge to nuclear science and technology ... When a manned base on the Moon is established, nuclear reactors will be the basic power supply for all that colony's activities.[13]

Starting in the late 1950s, American engineers designed a series of miniature reactors with the explicit aim of flying them into space. One of them, a prototype called SNAP-10A, actually made it.

In April 1965, SNAP-10A left Earth aboard a rocket and slipped into orbit 800 miles above the ground. Engineers meticulously designed it to be as small and as light as possible. It wasn't much bigger than a microwave. Controllable beryllium reflectors bounced stray neutrons back into its highly enriched uranium core, and an infusion of zirconium hydride moderated the neutrons. The whole thing was cooled by the same liquid metal mixture of sodium ('Na') and potassium ('K') – 'knack' – we find in fast reactors here on Earth. SNAP-10A dutifully converted its nuclear heat into 500 watts of electricity using the Seebeck effect, but a fault cut the mission short 43 days after launch. The failure lay with an electrical component; the nuclear reactor worked flawlessly, even in zero-gravity and after the trauma of launch, which is even more impressive considering SNAP-10A was the first nuclear reactor flown in space. It's still in orbit and will circle Earth for at least 4,000 years, after which time the fission fragments will have decayed into nothingness.[14]

Meanwhile, the USSR developed its own miniature reactors. Officials said the exploration of outer space was 'unthinkable without the use of nuclear power sources for thermal and electrical energy', and the Soviets launched their first space reactor in 1970. It generated 3,000 watts of electricity and completed one orbit around the Earth. In the latter half of the 1980s, the USSR put at least 10 reactors in orbit, one of which generated 6,000 watts of electricity and circled the Earth for 49 weeks.[15]

Nobody has landed a nuclear reactor on the surface of another world. Nobody has even flown one in space since the USSR in March 1988. Like so much nuclear technology, space reactors are vintage inventions that were never fully realised. Imagine what's possible now, with modern launch systems and after decades of technological advancement.

Thanks to renewed interest from the world's space agencies in returning to the Moon and (eventually) visiting Mars, space reactors are making their way back into the astronomic conversation. The US Department of Energy explicitly advocates for miniature nuclear reactors for long-duration missions throughout the solar system in its 2021 'Energy for Space' strategy. The Russian space agency and China National Space Administration intend to build a collaborative nuclear reactor on the Moon – flown there aboard a nuclear-powered cargo rocket – by the mid-2030s. And in 2024, the UK Space Agency awarded £4.8 million to Rolls-Royce to design a megawatt-scale micro-reactor that will generate electricity on the lunar surface.[16] We're on the brink of the Second Space Race, but this time with a nuclear twist.

Another giant leap

Earth is the biggest rock in the solar system. It takes a tremendous amount of energy to climb out of its deep gravitational well to orbit, and even more energy to fly to the Moon and beyond. Energy, as it is here on the ground, is the key to life in space.

Chemical rockets generate thrust by combusting fuel and directing the roaring jet through a nozzle. Hot gas goes one way, and Newton's Third Law – 'every action has an equal and opposite reaction' – pushes the rocket the other. Ever since a chemical rocket launched Sputnik into space in 1957, they've allowed humanity to deploy constellations of satellites in Earth's orbit, fling probes to all reaches of the solar system, and transport humans safely to the Moon (and back again).

Whilst they're impressive feats of engineering, chemical rockets have more or less hit the upper limit of their performance. Sure, there are gains to be made in cost and safety, and we can build ever bigger rockets. But chemical rockets are limited by . . . chemistry. Chemical fuel is energy-dilute. A chemical rocket must burn a *lot* of it to eke out a comparatively tiny quantity of energy. The rocket also must carry the hefty fuel with it, and then take *more* fuel to carry the fuel, and then take *even more* fuel to carry *that* fuel . . . and so on. This vicious cycle conspires to make chemical rockets incredibly weighty and inefficient. Alas, we won't see giant leaps forward in their speeds, performance, ranges, or payloads.

Chemical rockets are great at launching spacecraft a few hundred miles into Earth's orbit, but they're not so great at going further afield. This poses a dilemma for the human exploration of Mars (and beyond). During a 9-month journey to the Red Planet, astronauts would be exposed to about 500 millisieverts of cosmic radiation, or about 200 years' worth of terrestrial background radiation. They'd get a comparable dose on the way back. An unexpected solar flare would push the dose even higher. And once they arrived at Mars, they'd have to loiter on the surface until the planets swung themselves into the optimal alignment for the return trip, which would take more than a year. In all, the round trip would exceed 3 years, and an unacceptable radiation burden would be placed on a human crew. There are the deleterious psychological impacts of being cooped aboard a spaceship for months on end to consider, too. And whilst the mind withers from the isolation, the body would waste from the extended period of zero-gravity. Transporting people to Mars could well turn out to be a problem of mind and body rather than one of engineering.[17]

All this would be solved (or at least alleviated) by building faster rockets. Faster rockets demand propulsion systems that run on energy-dense fuel: *nuclear fuel.*

At their most basic, nuclear rockets look and work the same way as their chemical counterparts: they fire a jet of hot gas from a nozzle. The difference lies in the way they generate the heat. Instead of combusting chemical fuel, nuclear rockets coax energy from uranium-235 atoms in a miniature nuclear reactor, which they use to raise liquid hydrogen to scalding temperatures. The hydrogen boils rapidly, expands enormously, and shoots from a nozzle. Newton and his Third Law take care of the rest. And because uranium is millions of times more energy-dense than chemical fuel, nuclear rockets carry enormous amounts of energy in a small volume. They get more push from the same amount of fuel or, likewise, the same amount of push from less fuel. The *Saturn V* rocket that took the Apollo astronauts to the Moon burned through 2,500 tonnes of chemical fuel, which is why it was 360 feet tall. The same amount of energy is contained in less than 2 kilograms of highly enriched uranium.[18]

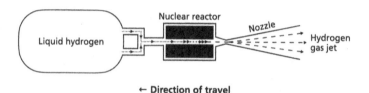

A nuclear rocket engine. Gas heated by fission goes one way; the rocket goes the other.

Nuclear rockets don't release their energy as quickly as chemical rockets, and so they're no good for getting off the ground. But once they're *in* space – if a chemical rocket carried one into orbit, say – they come into their own because they can fire for far longer. Chemical rockets coast to their destination once they're on their way. The practically inexhaustible supply of fuel carried by nuclear rockets, by contrast, lets them accelerate practically indefinitely; they can spend the first half of the journey speeding up and the second half slowing down. A 9-month crossing to Mars is cut almost in half. With the increased energy, nuclear rockets could take more direct routes to and from Mars, meaning astronauts wouldn't have to wait for the planets to line back up for the return leg. The radiation burden would lighten significantly; the psychological and physiological effects of space flight would be diminished. We could also take more cargo with us, like habitats, life-support systems, and micro-reactors. Nuclear rockets bring

Mars within touching distance, just as chemical rockets did with the Moon.[19]

The advantages of nuclear rockets were recognised as early as 1953, and by 1959, NASA had teamed up with the US Atomic Energy Commission to build a nuclear-powered rocket called *Kiwi* for long-duration journeys. Like the flightless bird after which it was named, the *Kiwi* rockets were destined never to fly. The flock was tested upside-down in Jackass Flats deep in the Nevada desert fixed firmly to scaffolding. And it worked! *Kiwi* after *Kiwi* blasted jet after jet of scalding hydrogen gas skyward from their nozzles, creating thrust that could push them through space. Through the 1960s and early 1970s, scientists tested dozens of rockets, each time fixing pitfalls and optimising performance. Some of them ran at a full 1,000 megawatts.

Alas, *Kiwi*'s descendants never flew in space. The US Government scrapped the nuclear rocket programme in 1973 before any of them made it off the ground. Little has happened since. Nuclear rockets are another example of vintage technology that never saw its full potential. The US Government spent about $3.5 billion on nuclear rocket research; by contrast, it spent just shy of $20 billion on Apollo. A relatively small group of dedicated scientists and engineers turned nuclear rockets from a dream into a reality in under 20 years on a relatively shoestring budget.[20]

NASA wants to send astronauts to Mars and bring them safely back home in the 2030s. To meet this end, it has rekindled its interest in nuclear rockets. In 2021, Associate Administrator of NASA's Science Mission Directorate Thomas Zurbuchen said, 'Future nuclear power and propulsion systems will . . . play a crucial role in enabling long-term human missions to the Moon and Mars.' NASA wants to fly one in space by the late 2020s and in 2023 committed $300 million to develop the *Demonstration Rocket for Agile Cislunar Operations* (*DRACO* for short). Let's not blow it this time.[21]

A future is possible in which astronauts voyage to Mars aboard nuclear-powered rockets, where they live in nuclear-powered habitats upon arrival, outside of which they sift and explore the red sands using nuclear battery-powered instruments. *Ad astra!*

The voyage goes on

So, what became of *Voyager 1* after its gravitational encounter with Titan in November 1980? The planetary phase of its mission may have ended, but the story was far from over.

Voyager 1 was deflected high above the solar system's flat orbital plane at more than 10 miles per second. Its onboard plutonium-238 continued to glow red-hot. With enough power to electrify its instruments, *Voyager 1* measured the waves and particles that stream through the distal regions of our solar system.

Thirty-two years later, on 25 August 2012, *Voyager 1* crossed the heliopause: it exited the bubble of the solar wind's influence and entered the space between the stars. *Voyager 2* crossed the heliopause on 5 November 2018. Both spacecraft continue to communicate with Earth.* [22]

But as their plutonium-238 decays and dulls, so does their electrical power. Engineers have remotely shut down their instruments one by one to preserve power, but any year now, the power will run too low, and both probes will go quiet. Sometime around 2036, they'll slip out of range of Earth's Deep Space Network and be lost forever. In the year 42,025 (or thereabouts), *Voyager 1* will pass within 1.6 light years of a star named AC79 3888 in the constellation of Camelopardalis; in the year 298,025, *Voyager 2* will pass within 4.3 light years of the brightest star in Earth's sky, Sirius the Dog Star.[23]

❈ ❈ ❈

Imagine we boarded a ship and set sail towards the constellation of Ophiuchus at light-speed. As we rush towards the stars at 186,000 miles per second, we glance over our shoulders at the receding shores

---

* As I write these words in the autumn of 2024, *Voyager 1* is 15.5 billion miles – 23 light-hours – from home; *Voyager 2* is 12.9 billion miles – more than 19 light-hours – from home. You can check the live status of the *Voyager* twins on NASA's Jet Propulsion Laboratory website.

of Earth. It's the last time we'll set eyes on home. In hardly any time at all, Earth becomes lost amongst a sea of stars.

After a day or so of sailing starward, the solar system is but a distant ripple on the ocean. We're destined for a long crossing in the space between the stars, and so we make ourselves comfortable . . . but a glimmer of gold straight ahead catches our eyes. It's *Voyager 1*, the most distant human-made artefact ever launched into space.

The lonely spacecraft is travelling at 38,000 miles per hour, quite a bit less hurried than our light-speed, so we slow down and pull up alongside. *Voyager*'s towering communication dish points quietly back at the pale blue dot whence it came, and its instruments and sensors protrude like masts. We spot its nuclear battery. And then we notice the source of the glimmer: a golden circle, 12 inches across, fixed to the side of *Voyager*'s keel. Inscribed into the disc is a series of pictographs, including a stamp triangulating the position of Earth and, curiously, instructions on how to play a vinyl record. On closer inspection, we realise the golden circle is actually a case, which we prise open. We find a golden 12-inch phonograph record inside. It even has a stylus. On the inside circle of the record's label runs the hand-etched message: TO THE MAKERS OF MUSIC – ALL WORLDS / ALL TIMES.

Following the guide, we drop the needle into the outermost groove, and after a pop and a crackle, the Golden Record fills the void of interstellar space with our finest art.

For 90 minutes, we listen to the sounds of Earth. It's an eclectic mix. Western classical – Beethoven, Mozart, Bach – Navajo flutes, Solomon Island panpipes, Aboriginal singing, and Peruvian song. Louis Armstrong and His Hot Seven jazzes some 'Melancholy Blues'. Blind Willie Johnson strums along with 'Dark Was the Night, Cold Was the Ground'. Chuck Berry rock 'n' rolls to 'Johnny B. Goode'.

And then the music gives way to the humdrum sounds of home. A dog barks. A steam train chugs by. Surf laps against the shoreline and thunder cracks in the sky. An elephant toots amongst chuckling hyenas. A heart beats. With a soft voice, a mother comforts her crying baby. They're such everyday sounds back home, but set amongst the every*thing* of the universe, they're miraculous.

And then we hear spoken words, and 55 people greet us in their native tongues. A woman says, 'Greetings to our friends in the stars, we wish that we will meet you someday' in Arabic. Another chirps, 'Hello everybody' in French. A man says 'Peace' in Hebrew.

Encoded in the record's grooves are images in analogue form. We raster the images across the vintage computer screen. They depict the extraordinary: a DNA double helix; the Earth as seen from orbit and the Moon's cratered face; an astronaut. They also depict the everyday: a young woman enjoying a bunch of grapes in a supermarket; daffodils; Olympic sprinters (with the USSR's Valeri Borzov in the lead); the silhouette of a human embryo; rush hour traffic in Thailand.

And then, the words of President Jimmy Carter scroll across the screen:

> This is a present from a small distant world, a token of our sounds, our science, our images, our music, our thoughts, and our feelings. We are attempting to survive our time so we may live into yours.

As we read, a familiar sound greets us. *Click click click*. Our Geiger counter senses the alpha radiation emanating from the record's case. There's no air our here to block it, which is why we can detect it at a distance. It's a thousandth of a gram of uranium-238 embedded in the gold. The smidgen contains some million trillion atoms. The scientists and engineers who crafted *Voyager* – master shipbuilders who believed in a time after their own – put it there as a timepiece: the uranium decays into non-radioactive lead-206 at a rate set by its half-life. By measuring the proportions of uranium-238 and lead-206 – with a mass spectrometer, say – the time elapsed since the record was made could be easily calculated. With a half-life of 4.5 billion years, the radioactive clock will tick for aeons.

We slip the Golden Record back in its case, hopeful that somewhere, sometime, someone will rediscover it and think well of us. Following the pulsar map, they'll triangulate *Voyager*'s origin in space: Earth. Measuring the uranium-238's radioactivity, they'll tell *Voyager*'s origin in time: the twentieth century.

In a world with increasing energy demands, it remains to be seen whether or not nuclear science will define humanity's planetary legacy. Either way, nuclear science has already defined our cosmic legacy. When the winds have worn the Pyramids to sand and the rains have washed the temples into the sea, the *Voyager* twins will carry the human spirit through the Milky Way. They might one day be the last surviving human artefacts, and if they are, then our science and art will be humanity's lasting legacy in the cosmos.

Epilogue: Appealing to Our Better Nature

When life allows, I spend my evenings and weekends amongst the deep valleys and high mountains of the Lake District. My soul is soothed when I hike, run, swim, cycle, birdwatch, and photograph. Alfred Wainwright describes it best in the seventh volume of his *Pictorial Guide to the Lakeland Fells*: 'For a man trying to get a persistent worry out of his mind, the top of Haystacks is a wonderful cure.' Earth's idyllic landscapes, and the flora and fauna that inhabit them, enchant us. We all cherish Darwin's living and breathing 'endless forms most beautiful and most wonderful'.

But nature conjures many a tempest, too: high infant mortality, famine, short lifespans, drought, infection, cancer. All these things are as natural as mountains such as Haystacks and ecosystems such as coral reefs. Ernest Becker writes of 'the callous unconcern of nature' in *The Denial of Death*. His description is apt: every other child died before their 15th birthday until healthcare began improving in the nineteenth century.[1] 'Nature is not cruel,' Richard Dawkins writes in *River Out of Eden*, 'only pitilessly indifferent.'

And this, I think, is where we let our fondness for the pleasant and aesthetic sides of nature cloud our perspective. Blinded by the idyll, there's a strong tendency to believe that 'natural' things are good, and 'unnatural' things are – by definition – bad. This *appeal to nature fallacy* is deeply misguided. It manifests in clichés such as 'living in harmony with nature' (does this mean living in harmony with beautiful landscapes or living in harmony with famine?) and 'untouched by human hands' (and this an untouched rainforest or an unstitched wound?). The fallacy also permeates our culture. Supermarkets exploit it by pricing up foods made from 'natural ingredients' that are no better for you than 'unnatural', selectively bred crops. Retailers promulgate it by advertising clothes made from 'natural fibres', as though they're automatically better than 'unnatural' textiles such as polyester and

Gore-Tex. Quacks promulgate it by peddling 'natural remedies', as though 'unnatural' medicines such as paracetamol and chemotherapy are inferior. And environmentalists exploit it by promoting sources of energy that *feel* natural, such as wind-blown turbines and sunshine-powered panels, whilst opposing those that feel unnatural, such as uranium-powered reactors.

In reality, there's nothing natural about a wind turbine or a solar panel; they're as high-tech as smartphones and hearing aids. And even if renewables *were* somehow more 'natural' than nuclear reactors, so what? The pinnacles of human achievement are defined by unnatural creations, such as feeding people with nitrogen fertilisers, connecting them with transistors, curing them with antibiotics, and sending them to space aboard rockets. Harvesting the clean and abundant energy our world so desperately needs by changing the atomic building blocks of matter is also on that list. All these things are 'unnatural' – *Homo sapiens* alone does them – but they're all also *good*.

And there's another misguided, often unspoken implication, of this fallacy: human civilisation, in all its unnaturalness, is a force of evil. This douses environmentalism with nihilism and hopelessness. What's the point in building a better future if there isn't a future worth saving? Does humanity even *deserve* a better future?

We shouldn't second guess ourselves. We *can* continue social and economic advancement, and we *can* preserve the Earth's splendours. But to do so, we must abandon our proclivity for leaning towards 'natural' solutions. Building nuclear power stations and circularising their fuel cycles with breeder reactors – starting now, and in earnest – is the single biggest environmental step we could take. It's the way we'll power the world – cleanly and reliably – far beyond our net zero deadlines; carbon dioxide doesn't enter the nuclear equation.

Nuclear science, implemented widely and to its full promise, represents a belief in a positive future. *Going Nuclear* is my case, as a nuclear environmentalist, for that future.

I'm going back to the lab now. Over to you.

Appendix

Numerical prefixes

We use numerical prefixes as mathematical shorthands. They can be bolted onto the front of any unit, including units of power (watts) and energy (joules, watt-hours, or electron-volts).

Prefix	Symbol	Multiplier		Everyday word
Pico	p	10^{-12}	= 0.000000000001	'Trillionth'
Nano	n	10^{-9}	= 0.000000001	'Billionth'
Micro	µ	10^{-6}	= 0.000001	'Millionth'
Milli	m	10^{-3}	= 0.001	'Thousandth'
—	—	10^{0}	= 1	—
Centi	c	10^{2}	= 100	'Hundred'
Kilo	k	10^{3}	= 1,000	'Thousand'
Mega	M	10^{6}	= 1,000,000	'Million'
Giga	G	10^{9}	= 1,000,000,000	'Billion'
Tera	T	10^{12}	= 1,000,000,000,000	'Trillion'

List of Illustrations

2 pp To Come

Bibliography

@bp_plc. 2019a. X.
@bp_plc. 2019b. X.
@GoldGooseAward. nd. X.
@Shell_NatGas. 2018. X.
350.org. 2023. 350.org Responds to Announcement By Countries to Triple Nuclear Energy.
Aaltonen et al. 2016. Geology of Olkiluoto. Posiva Report.
Advancing Nuclear Medicine. nd. Yttrium-90.
Agency for Natural Resources and Energy. 2021. Outline of Strategic Energy Plan.
Ahenkorah et al. 2021. Bismuth-213 for Targeted Radionuclide Therapy: From Atom to Bedside. Pharmaceutics.
Al-Ibraheem and Scott. 2023. 161Tb-PSMA Unleashed: a Promising New Player in the Theranostics of Prostate Cancer. Nuclear Medicine and Molecular Imaging.
Albertsson et al. 2023. Astatine-211 based radionuclide therapy: Current clinical trial landscape. Frontiers in Medicine.
Alvarez. 2013. Managing the Uranium-233 Stockpile of the United States. Science & Global Security.
American Academy of Achievement. 2016. Glenn T. Seaborg, Academy Class of 1972, Full Interview. YouTube.
American Institute of Physics. 1967. Otto Frisch, Oral History.
American Institute of Physics. 2007. Discovery of Fission.
American Institute of Physics. 2020. Ernest Moniz, oral history.
American Nuclear Society. 2022. Another delay, cost bump, for Flamanville-3.
American Nuclear Society. 2024a. Sweden's SKB approved to begin construction of spent fuel repository.
American Nuclear Society. 2024b. Nations envision nuclear reactors on the moon.
Anderson. 1966. Prof. S. K. Allison. Nature.

ANSTO. nd. Radioactive capsule goes missing in Australian outback.

Apostolidis et al. 2005. Cyclotron production of Ac-225 for targeted alpha therapy. Applied Radiation and Isotopes.

Arms Control Association. 2023. The Nuclear Testing Tally.

Arms Control Association. nd. U.S.-Russian Nuclear Arms Control Agreements at a Glance.

Atomic Heritage Foundation. 2014. Tsar Bomba.

Atomic Heritage Foundation. 2018. Nuclear Close Calls: The Norwegian Rocket Incident.

Atomic Heritage Foundation. 2022a. Lise Meitner.

Atomic Heritage Foundation. 2022b. Einstein-Szilard Letter.

Atomic Heritage Foundation. nd. Marie Curie.

Australian Government. 2023. Environment Protection and Biodiversity Conservation Act 1999, C2004A00485.

Ball. 2004. How Radioactive Is Your Banana? Journal of Chemical Education.

Baumhover. 2002. A Personal Account of Developing the Sterile Insect Technique to Eradicate the Screwworm from Curacao, Florida and the Southeastern United States. Florida Entomologist. DOI: 10.1653/0015-4040(2002)0850666:APAODT2.0.CO;2.

BBC Focus. 2023. Finland's plan to bury spent nuclear fuel for 100,000 years.

BBC Future. 2023. What are the odds of a successful space launch?

BBC News. 2015a. Poaching the creature that's more valuable than gold.

BBC News. 2015b. Litvinenko: A deadly trail of polonium.

BBC News. 2016. Litvinenko suspects Andrei Lugovoi and Dmitry Kovtun.

BBC News. 2017. Poachers kill rhino for his horn at French zoo.

BBC News. 2018. Japan confirms first Fukushima worker death from radiation.

BBC News. 2019. Greta Thunberg, climate change activist, sails into New York City.

BBC News. 2020a. UK can be 'Saudi Arabia of wind power' – PM.

BBC News. 2020b. Boris Johnson: Wind farms could power every home by 2030.

BBC News. 2022a. Energy strategy: No plans for new nuclear sites in Scotland.

BBC News. 2022b. Cottingham: Europe's biggest battery storage system switched on.
BBC News. 2022c. Breakthrough in nuclear fusion energy announced.
BBC News. 2023a. Emergency coal power plants used for first time as UK sees cold snap.
BBC News. 2023b. UK weather: Ratcliffe-on-Soar power station readied to boost supply.
BBC News. 2023c. Germans split as last three nuclear power stations go off grid.
BBC News. 2023d. EDF: French energy giant posts worst-ever results.
BBC News. 2023e. The science behind the Fukushima waste water release.
BBC News. 2023f. Fukushima: China retaliates as Japan releases treated nuclear water.
BBC News. 2023g. Rio Tinto apologises for losing radioactive capsule in Australia.
BBC News. 2023h. How a tiny radioactive capsule was found in Australia's vast outback.
BBC News. 2024a. Microsoft chooses infamous nuclear site for AI power.
BBC News. 2024b. Google turns to nuclear to power AI data centres.
BBC News. 2024c. Is nuclear power gaining new energy?
BBC News. 2024d. Drax: UK power station still burning rare forest wood.
Becquerel. 1896a. Sur les radiations émises par phosphorescence. Comptes rendus de l'Académie des Sciences.
Becquerel. 1896b. Sur les radiations invisibles émises par les corps phosphorescents. Comptes rendus de l'Académie des Sciences.
Benešová *et al.* 2015. Preclinical Evaluation of a Tailor-Made DOTA-Conjugated PSMA Inhibitor with Optimized Linker Moiety for Imaging and Endoradiotherapy of Prostate Cancer. Journal of Nuclear Medicine.
Bennett *et al.* 2006. Health effects of the Chernobyl accident and special health care programmes: report of the UN Chernobyl Forum Expert Group "Health". World Health Organization. ISBN: 978-92-4-159417-2.
Bennett *et al.* 2016. I-131 Therapy for Thyroid Cancer. Section 12: Nuclear medicine therapy in Diagnostic Imaging: Nuclear Medicine. ISBN: 978-0-323-37753-9.
Bhabha. 1948. Note on the Organisation of Atomic Research in India. Board of Research on Atomic Energy.

Black. 2012. Witness statement of Stuart Black (INQ014291). The Litvinenko Inquiry (The National Archives).

Blackett. 1925. The ejection of protons from nitrogen nuclei, photographed by the Wilson method. Proceedings of the Royal Society of London. Series A, Containing Papers of a Mathematical and Physical Character.

Bleek. 2017. When Did (and Didn't) States Proliferate? Chronicling the Spread of Nuclear Weapons (Discussion Paper). Cambridge, MA: Project on Managing the Atom, Belfer Center for Science and International Affairs, Harvard Kennedy School and the James Martin Center for Nonproliferation Studies, Middlebury Institute of International Studies, Monterey, CA.

Bloomberg. 2023a. Germany Fires Up Extra Coal Unit to Help in Cold Snap.

Bloomberg. 2023b. Nuclear Power Makes Comeback with Massive Carbon-Free Vogtle Plant in Georgia.

Bodel et al. 2023. Managing the UK plutonium stockpile: no easy choices. Dalton Nuclear Institute.

Boice. 2015. The Boice Report #40. Health Physics News.

Boiteux. 2009. Le programme électro-nucléaire: EDF et ses choix industriels in État et énergie XIXe-XXe siècle. ISBN: 978-2-11-129456-1.

Breeze. 2017. Chapter 4: Water-Cooled Reactors in Nuclear Power. Elsevier. ISBN: 978-0-08-101043-3.

Bretscher and Cockcroft. 1955. Enrico Fermi, 1901–1954. Biographical Memoirs of Fellows of the Royal Society.

British Antarctic Survey. 2018. Towards Net Zero Carbon.

Bromet et al. 2011. A 25 Year Retrospective Review of the Psychological Consequences of the Chernobyl Accident. Clinical Oncology.

Brown et al. 2018. Americium and Plutonium Purification by Extraction (the AMPPEX process): Development of a new method to separate 241Am from aged plutonium dioxide for use in space power systems. Progress in Nuclear Energy.

Bushland and Hopkins. 1953. Sterilization of Screw-Worm Flies with X-Rays and Gamma-Rays. Journal of Economic Entomology.

Calcagnile et al. 2013. Determining 14C Content in Different Human Tissues: Implications for Application of 14C Bomb-Spike Dating in Forensic Medicine. Radiocarbon.

Carbajo. 2001. A review of the thermophysical properties of MOX and UO_2 fuels. Journal of Nuclear Materials.

Centres for Disease Control and Prevention. 2019. Acute Radiation Syndrome: A Fact Sheet for Clinicians.

Centro de Dispersión de Moscas Estériles. nd. Processes.

China Nuclear Energy Association. 2021. China Nuclear Energy Yearbook.

Choppin et al. 2013a. Chapter 21: The Nuclear Fuel Cycle in Radiochemistry and Nuclear Chemistry (Fourth Edition). Elsevier. ISBN: 978-0-12-405897-2.

Choppin et al. 2013b. Cosmic Radiation and Radioelements in Nature in Radiochemistry and Nuclear Chemistry (Fourth Edition). Elsevier. ISBN: 978-0-12-405897-2.

Clean Energy Wire. 2017. Quotes from the Berlin Energy Transition Dialogue.

CNBC. 2020. Joe Biden calls climate change the 'number one issue facing humanity'.

Cochran et al. 2010. Fast Breeder Reactor Programs: History and Status. International Panel on Fissile Materials. ISBN: 978-0-9819275-6-5.

Cole et al. 2014. Radiosyntheses using Fluorine-18: The Art and Science of Late Stage Fluorination. Current Topics in Medicinal Chemistry.

Committee on Climate Change. 2020. Sixth Carbon Budget.

Coppedge et al. 2024. V-Dem Country-Year Dataset v14.

Cucinotta et al. 2008. Physical and Biological Organ Dosimetry Analysis for International Space Station Astronauts. Radiation Research.

Curie. 1898. Rayons émis par les composés de l'uranium et du thorium. Comptes Rendus.

Curie. 1905. Radioactive substances, especially radium (Nobel Lecture, June 6, 1905). The Nobel Prize.

Curry. 2016. Evolution Made to Order: Plant Breeding and Technological Innovation in Twentieth-Century America. University of Chicago Press. ISBN: 978-0226790862.

Czernin and Calais. 2021. 177Lu-PSMA617 and the VISION Trial: One of the Greatest Success Stories in the History of Nuclear Medicine. Journal of Nuclear Medicine.

Dattani et al. 2023a. Life Expectancy. Our World in Data.

Dattani et al. 2023b. Child and Infant Mortality. Our World in Data.

DelCul and Spencer. 2020. Reprocessing and recycling in Advances in Nuclear Fuel Chemistry. Woodhead Publishing. ISBN: 978-0-08-102571-0.

Department for Business, Energy, and Industrial Strategy. 2021. Energy Trends: December 2021.

Department for Business, Energy, and Industrial Strategy. 2023. Advanced Nuclear Technologies.

Department for Energy Security and Net Zero. 2024a. Domestic electricity prices in the EU for small, medium and large consumers.

Department for Energy Security and Net Zero. 2024b. Domestic gas prices in the EU for small, medium and large consumers.

Department for Energy Security and Net Zero. 2024c Renewable electricity capacity and generation (ET 6.1 - quarterly) in Energy Trends: UK renewables.

Department for Energy Security and Net Zero. 2024d. Sub-national electricity consumption statistics 2005 to 2022. Regional and local authority electricity consumption statistics.

Department for Energy Security and Net Zero. 2024e Civil Nuclear: Roadmap to 2050. ISBN: 978-1-5286-4560-7.

Department of Energy & Climate Change. 2013. Hinkley Point C.

Department of the Air Force. 1985. Operation Ivy. Project 7. 5. Dispersion of gaseous debris from nuclear explosions (ADA995448).

Der Spiegel. 2019. German Failure on the Road to a Renewable Future.

Deryabina *et al.* 2015. Long-term census data reveal abundant wildlife populations at Chernobyl.

Diehl *et al.* 2023. Peace data: Concept, measurement, patterns, and research agenda. Conflict Management and Peace Science.

Dombey. 2007. Report by Professor Norman Dombey (INQ006067). The Litvinenko Inquiry (The National Archives).

Duderstadt and Hamilton. 1976a. Chapter 3: Fission Chain Reactions and Nuclear Reactors - an introduction in Nuclear Reactor Analysis. ISBN: 978-0-471-22363-4.

Duderstadt and Hamilton. 1976b. Chapter 2: The Nuclear Physics of Fission Chain Reactions in Nuclear Reactor Analysis. ISBN: 978-0-471-22363-4.

Dustin and Borrelli. 2021. Assessment of alternative radionuclides for use in a radioisotope thermoelectric generator. Nuclear Engineering and Design.

EDF. 2019. All you need to know about wind power.

EDF. 2020. A guide to solar panels.

EDF. nd. How much it costs to charge & run an electric car.

El-Showk. 2022. Finland built this tomb to store nuclear waste. Can it survive for 100,000 years? Science.

ElBaradei. 2009. Taking control of nuclear. The Guardian.

Embassy of the People's Republic of China in the Kingdom of Thailand. 2023. Foreign Ministry Spokesperson's Statement on the Japanese Government's Start of Releasing Fukushima Nuclear-Contaminated Water into the Ocean.

Ember. 2024. Yearly electricity data.

Emblemsvåg. 2024. What if Germany had invested in nuclear power? A comparison between the German energy policy the last 20 years and an alternative policy of investing in nuclear power. International Journal of Sustainable Energy.

Emirates Nuclear Energy Corporation. 2024. UAE Celebrates Historic Milestone as Unit 4 of the Barakah Plant Commences Commercial Operation: Delivering Full-fleet Operations.

Emrich. 2022. Promises and Challenges of Nuclear Propulsion for Space Travel. International Atomic Energy Agency 'Atoms for Space: Nuclear Systems for Space Exploration' webinar.

Energy Information Administration. 2023. International Energy Data.

Energy Institute. 2024. Statistical Review of World Energy.

Energy Voice. 2022. National Grid puts coal power on standby to meet UK cold snap.

Erisman *et al.* 2008. How a century of ammonia synthesis changed the world. Nature Geoscience.

Euractiv. 2023a. Seven countries reject nuclear-derived hydrogen from EU renewables law.

Euractiv. 2023b. Paris drafts European 'nuclear alliance'.

Euractiv. 2023c. Macron at COP28: 'Nuclear energy is back'.

Euronews. 2023a. Austria launches legal challenge over EU's 'greenwashing' of nuclear and gas.

Euronews. 2023b. Video. Greenpeace celebrates end of Germany's nuclear era with T.Rex dinosaur.

European Commission, Joint Research Centre. 2019. European atlas of natural radiation.

European Commission. 2020. Proposal for a Council Decision on the Adoption of the 2020-2023 High Flux Reactor Supplementary Research Programme at Petten to be implemented by the Joint Research Centre for the European Atomic Energy Community (Document 52020PC0108).

European Court of Human Rights. 2021. Russia was responsible for assassination of Aleksandr Litvinenko in the UK (press release).

European Medicines Agency. 2022. Pluvicto.

European Nuclear Society. nd. Pressure vessel.

European Parliament. 2021. Regulation (EU) 2021/1119 of the European Parliament and of the Council of 30 June 2021 establishing the framework for achieving climate neutrality and amending Regulations (EC) No 401/2009 and (EU) 2018/1999 ('European Climate Law').

European Parliament. 2023. Annex IV, European Union Emissions Trading SystemDirective 2003/87/EC.

Eurostat. 2023. Cancer statistics.

Eurostat. 2024a. Passenger cars in the EU.

Eurostat. 2024b. Electricity prices for household consumers, nrg_PC_204.

Eve. 1939. Rutherford: Being the Life of Letters of the Rt Hon. Lord Rutherford, O. M. Cambridge at the University Press.

Fallah *et al.* 2023. FDA Approval Summary: Lutetium Lu 177 Vipivotide Tetraxetan for Patients with Metastatic Castration-Resistant Prostate Cancer. Clinical Cancer Research.

Fan *et al.* 2010. Safety Assessment on Nue Fuel Bundles for Demonstration Irradiation in a CANDU Reactor. 11th International Conference on CANDU Fuel.

Fauna & Flora International. nd,a. Northern white rhino.

Fauna & Flora International. nd,b. Pangolins.

Federal Ministry of Digital and Transport. 2023. Fahrleistungserhebung (FLE).

Fermi. 1934. Possible Production of Elements of Atomic Number Higher than 92. Nature.

Fermi. 1945. My Observations During the Explosion at Trinity on July 16, 1945.

Fermi. 1952. Experimental Production of a Divergent Chain Reaction. U. S. Department of Energy.

Ferris *et al.* 2021. Use of radioiodine in nuclear medicine—A brief overview. Journal of Labelled Compounds and Radiopharmaceuticals.

Financial Times. 2014. German plea to Sweden over threat to coal mines.

Financial Times. 2018. Fukushima nuclear disaster: did the evacuation raise the death toll?.

Financial Times. 2021. Japan says nuclear crucial to hitting net zero goal by 2050.

Finlex. 2020. Nuclear Energy Act (990/1987).

Finnish Energy. 2022. Popularity of nuclear power reaches a record high in Finland.

Flyvbjerg. 2013. Over Budget, Over Time, Over and Over Again: Managing Major Projects in The Oxford Handbook of Project Management. ISBN: 978-0-19-172487-9.

Forbes. 2024. Photo Of Northern White Rhino Being Protected By Armed Guard Went Viral.

Foreign, Commonwealth Development Office. 2023. Statement on Iranian nuclear steps reported by the IAEA.

Fournier and Ross. 2013. Radiocarbon Dating: Implications for Establishing a Forensic Context. Forensic Science Policy & Management.

France 24. 2019. CO2 row over climate activist Thunberg's yacht trip to New York.

Franck. 1945. Memorandum on 'Political and Social Problems' from Members of the 'Metallurgical Laboratory' of the University of Chicago. Metallurgical Laboratory.

Friedlingstein *et al.* 2023a. Global Carbon Budget 2023. Earth System Science Data.

Friedlingstein *et al.* 2023b Supplemental data of Global Carbon Budget 2023. Global Carbon Project.

Friends of the Earth. 2024. Should we use nuclear energy?

Frisch. 1974. 'Somebody Turned the Sun on with a Switch'. Bulletin of the Atomic Scientists.

Fukushima Prefecture and International Atomic Energy Agency. 2021. Radiation Monitoring and Remediation Following the Fukushima Daiichi Nuclear Power Plant Accident.

Fukushima Prefecture. 2024. Fukushima Prefecture Radioactivity measurement map.

Gangopadhyay et al. 2006. Triggering radiation alarms after radioiodine treatment. BMJ.

Gasparini and Wasser. nd. Water & Ices on the Moon. NASA.

Geological Survey of Finland. 2023. Bedrock of Finland.

Geth et al. 2015. An overview of large-scale stationary electricity storage plants in Europe: Current status and new developments. Renewable and Sustainable Energy Reviews.

Gibney. 2024. Mars rover mission will use pioneering nuclear power source. Nature.

Giovannitti and Freed. 1965. The Decision to Drop the Bomb. Coward-McCann, Inc.

Gowlett. 2016. The discovery of fire by humans: a long and convoluted process. Philosophical Transactions of the Royal Society B.

Greenpeace. 2023. Japan announces date for Fukushima radioactive water release.

Greenpeace. nd. Nuclear energy.

Grenèche. 2002. French plutonium management program (IAEA-TECDOC--1286). International Atomic Energy Agency.

Grigsby et al. 2000. Radiation Exposure From Outpatient Radioactive Iodine (131I) Therapy for Thyroid Carcinoma. JAMA.

Grimm. 2008. The Mushroom Cloud's Silver Lining. Science.

Grolund. 2011. How Many Cancers Did Chernobyl Really Cause?– Updated Version. Union of Concerned Scientists.

Grubler. 2010. The costs of the French nuclear scale-up: A case of negative learning by doing. Energy Policy.

Guardian News. 2019. Greta Thunberg to world leaders: 'How dare you? You have stolen my dreams and my childhood'. YouTube.

Guardian News. 2023. 'The era of global boiling has arrived' warns the UN. YouTube.

Guerra Liberal et al. 2016. Palliative treatment of metastatic bone pain with radiopharmaceuticals: A perspective beyond Strontium-89 and Samarium-153. Applied Radiation and Isotopes.

Hagemann et al. 1970. Absolute isotopic scale for deuterium analysis of natural waters. Absolute D/H ratio for SMOW. Tellus A: Dynamic Meteorology and Oceanography.

Hahn and Strassmann. 1939. Über den Nachweis und das Verhalten der bei der Bestrahlung des Urans mittels Neutronen entstehenden Erdalkalimetalle. Naturwissenschaften.

Hall. 2023. NASA, DARPA Partner with Industry on Mars Rocket Engine. NASA.

Handl. 1998. 75 MW heat extraction from Beznau nuclear power plant. International Atomic Energy Agency.

Harrison and Fox. 2023. Biomass plant is UK's top emitter. Ember.

Harrison et al. 2007. Polonium-210 as a poison. Journal of Radiological Protection.

Harrison. 2021. UK biomass emits more CO_2 than coal. Ember.

HBO 2019. Episode 5: Vichnaya Pamyat. Chernobyl.

Health and Safety Executive. nd. Work-related fatal injuries in Great Britain, 2023.

Hecker. 2000. Plutonium and Its Alloys: From atoms to microstructure. Los Alamos Science.

Helal and Dadachova. 2018. Radioimmunotherapy as a Novel Approach in HIV, Bacterial, and Fungal Infectious Diseases. Cancer Biotherapy and Radiopharmaceuticals.

Hempelmann et al. 1952. The acute radiation syndrome: a study of nine cases and a review of the problem. Annals of Internal Medicine.

Hempelmann et al. 1979. What Has Happened to the Survivors of the Early Los Alamos Nuclear Accidents? Conference for Radiation Accident PreparednessOak Ridge, TN.

Herre et al. 2024a. Number of countries that approve of nuclear weapons treaties. Our World in Data.

Herre et al. 2024b. War and Peace. Our World in Data.

Hess et al. 1976. Preliminary Meteorological Results on Mars from the Viking 1 Lander. Science.

Hey and Walters. 1997. Chapter 7: Little Boy and Fat Man: relativity in action in Einstein's Mirror. Cambridge University Press. ISBN: 978-0-521-43532-1.

Hickman et al. 2021. Climate anxiety in children and young people and their beliefs about government responses to climate change: a global survey. The Lancet Planetary Health.

HM Government. 2019. Climate Change Act 2008.

Holden et al. 2018. IUPAC Periodic Table of the Elements and Isotopes (IPTEI) for the Education Community (IUPAC Technical Report). Pure and Applied Chemistry.

Hopkin. 2005. Carbon in teeth helps to identify disaster victims. Nature.

Hua et al. 2022. Atmospheric Radiocarbon for the Period 1950–2019. Radiocarbon.

Hula. 2015. Atomic Power in Space II: A history of space nuclear power and propulsion in the United States. Idaho National Laboratory.

Hyatt. 2017. Plutonium management policy in the United Kingdom: The need for a dual track strategy. Energy Policy.

Hyatt. 2020. Safe management of the UK separated plutonium inventory: a challenge of materials degradation. npj Materials Degradation.

Idaho National Engineering Laboratory. 1979. Experimental Breeder Reactor I. The American Society of Mechanical Engineers.

Idaho National Laboratory. 2019. EBR-I lights up the history of nuclear energy development.

Idel. 2022. Levelized Full System Costs of Electricity. Energy.

Independent. 2024. Last coal-fired power station in Britain closes down in landmark moment for clean energy.

Institut National d'Etudes Démographiques. 2020. Life expectancy in France.

Institute for Health Metrics and Evaluation. 2019. Air pollution.

International Atomic Energy Agency. 1989. Statute, as amended up to 28 December 1989.

International Atomic Energy Agency. 2001. IAEA Releases Nuclear Power Statistics for 2000.

International Atomic Energy Agency. 2005. The role of nuclear power and nuclear propulsion in the peaceful exploration of space. ISBN: 978-92-0-107404-1.

International Atomic Energy Agency. 2007. Management of Reprocessed Uranium: Current Status and Future Prospects. ISBN: 92-0-114506-3.

International Atomic Energy Agency. 2012a. Status of Fast Reactor Research and Technology Development. ISBN: 978-92-0-130610-4.

International Atomic Energy Agency. 2012b. Improved Barley Varieties - Feeding People from the Equator to the Arctic.

International Atomic Energy Agency. 2017. Bangladesh Triples Rice Production with Help of Nuclear Science.

International Atomic Energy Agency. 2019a. Status Report – UK SMR (Rolls-Royce and Partners) United Kingdom. Advanced Reactor Information System.

International Atomic Energy Agency. 2019b. France's Efficiency in the Nuclear Fuel Cycle: What Can 'Oui' Learn?

International Atomic Energy Agency. 2019c. World Thorium Occurrences, Deposits and Resources. ISBN: 978-92-0-103719-0.

International Atomic Energy Agency. 2020a. IAEA Releases 2019 Data on Nuclear Power Plants Operating Experience.

International Atomic Energy Agency. 2020b. Nuclear Technology Review 2020.

International Atomic Energy Agency. 2022a. Japan, Czech Republic Latest Countries to Join Forum Dedicated to Safe and Secure Deployment of SMRs.

International Atomic Energy Agency. 2022b. What is Nuclear Fusion?.

International Atomic Energy Agency. 2022c. Status and Trends in Spent Fuel and Radioactive Waste Management. ISBN: 978-92-0-130521-3.

International Atomic Energy Agency. 2022d. Near Term and Promising Long Term Options for the Deployment of Thorium Based Nuclear Energy.

International Atomic Energy Agency. 2022e. Fuji Nijo II. Mutant Variety Database.

International Atomic Energy Agency. 2022f. Kyoryoku-reikou. Mutant Variety Database.

International Atomic Energy Agency. 2022g. CC66/4. Mutant Variety Databse.

International Atomic Energy Agency. 2022h. Pink Hat. Mutant Variety Database.

International Atomic Energy Agency. 2022i. Search: tulip. Mutant Variety Database.

International Atomic Energy Agency. 2022j. Shobha. Mutant Variety Database.

International Atomic Energy Agency. 2022k. UNA-La Molina 95.

International Atomic Energy Agency. 2022l. Binadhan-7.

International Atomic Energy Agency. 2022m. Communication Received from the United Kingdom of Great Britain and Northern Ireland Concerning Its Policies Regarding the Management of Plutonium.

International Atomic Energy Agency. 2023a. List of Member States.

International Atomic Energy Agency. 2023b. Energy, Electricity and Nuclear Power Estimates for the Period up to 2050. ISBN: 978-92-0-137323-6.

International Atomic Energy Agency. 2023c. Spent Fuel and Radioactive Waste Information System: Finland.

International Atomic Energy Agency. 2023d. Communication Received from the United Kingdom of Great Britain and Northern Ireland Concerning Its Policies Regarding the Management of Plutonium.

International Atomic Energy Agency. 2023e. Communication Received from France Concerning Its Policies Regarding the Management of Plutonium.

International Atomic Energy Agency. 2023f. Comprehensive Report on the Safety Review of the ALPS-Treated Water at the Fukushima Daiichi Nuclear Power Station.

International Atomic Energy Agency. 2023g. IAEA Releases Annual Data on Illicit Trafficking of Nuclear and other Radioactive Material.

International Atomic Energy Agency. 2024a. Power Reactor Information System.

International Atomic Energy Agency. 2024b. Radiation protection of patients during PET/CT scanning.

International Atomic Energy Agency. 2024c Production and Quality Control of Actinium-225 Radiopharmaceuticals. International Atomic Energy Agency. ISBN: 978-92-0-121324-2.

International Atomic Energy Agency. 2024d. Incident and Trafficking Database 2024 Factsheet.

International Atomic Energy Agency. nd,a. Atoms for Peace Speech.

International Atomic Energy Agency. nd,b. Nuclear Fuel Cycle Database.

International Atomic Energy Agency. nd,c. Chernobyl, Frequently Asked Chernobyl Questions.

International Atomic Energy Agency. nd,d. Radiation in Everyday Life.

International Atomic Energy Agency. nd,e. Fukushima Daiichi ALPS Treated Water Discharge – FAQs.

International Atomic Energy Agency. nd,f. Medical sterilization.

International Atomic Energy Agency. nd,g. Mutant Variety Database.
International Atomic Energy Agency. nd,h. Tephritid fruit flies.
International Atomic Energy Agency. nd,i. Lepidoptera (moth pests).
International Atomic Energy Agency. nd,j. Malaria.
International Commission on Radiological Protection. 2007. The 2007 Recommendations of the International Commission on Radiological Protection. Elsevier. ISBN: 978-0-7020-3048-2.
International Energy Agency and OECD Nuclear Energy Agency. 2020. Projected Costs of Generating Electricity 2020.
International Energy Agency. 2021. Ammonia Technology Roadmap.
International Energy Agency. 2022a. Solar PV Global Supply Chains.
International Energy Agency. 2022b. France 2030 Investment Plan – Small Modular Reactor investment, Policies.
International Energy Agency. 2022c. The Role of Critical Minerals in Clean Energy Transitions.
International Energy Agency. 2023a. Energy Statistics Data Browser.
International Energy Agency. 2023b. Global Hydrogen Review 2023.
International Energy Agency. 2023c. World Energy Outlook 2023.
International Energy Agency. 2023d. Unit Converter.
International Energy Agency. 2024. World Energy Investment 2024.
International Energy Agency. nd,a. Buildings.
International Energy Agency. nd,b. Heating.
International Energy Agency. nd,c. Transport.
International Energy Agency. nd,d. Industry.
International Energy Agency. nd,e. Heat Pumps.
International Energy Agency. nd,f. Steel.
International Energy Agency. nd,g. Germany.
International Renewable Energy Agency. 2020. Green Hydrogen Cost Reduction: Scaling up Electrolysers to Meet the 1.5°C Climate Goal. ISBN: 978-92-9260-295-6.
International Renewable Energy Agency. 2022. Renewable Power Generation Costs in 2022.
International Renewable Energy Agency. 2024. Renewable Power Generation Costs in 2023. ISBN: 978-92-9260-621-3.
Jalilian and Albon. 2023. Neutrons Save Lives. International Atomic Energy Agency.

Jarvis et al. 2022. The Private and External Costs of Germany's Nuclear Phase-Out. Journal of the European Economic Association.

Jet Propulsion Laboratory. nd,a. Interstellar Mission.

Jet Propulsion Laboratory. nd,b. Frequently Asked Questions.

Johnson et al. 1967. Design, ground test and flight test of SNAP 10A, first reactor in space. Nuclear Engineering and Design.

Johnson. 2012. Safeguarding the atom: the nuclear enthusiasm of Muriel Howorth. The British Journal for the History of Science.

Joliot and Curie. 1934. Artificial Production of a New Kind of Radio-Element. Nature.

Joliot. 1935. Nobel Lecture: Chemical evidence of the transmutation of elements. The Nobel Prize.

Joyce. 2018. Chapter 10: Mainstream Power Reactor Systems in Nuclear Engineering. Elsevier. 978-0-08-100962-8.

Juwi. nd. PV power plant Kozani.

Ketchum. 1987. Lessons of Chernobyl: SNM Members Try to Decontaminate World Threatened by Fallout. Journal of Nuclear Medicine.

Kharecha and Hansen. 2013. Prevented Mortality and Greenhouse Gas Emissions from Historical and Projected Nuclear Power. Environmental Science & Technology.

Kharecha and Sato. 2019. Implications of energy and CO_2 emission changes in Japan and Germany after the Fukushima accident. Energy Policy.

Kittel et al. 1957. The EBR-1 Meltdown-Physical and Metallurgical Changes in the Core (Final Report - Metallurgy Program 7.2.18). Argonne National Laboratory.

Klaassen et al. 2019. The various therapeutic applications of the medical isotope holmium-166: a narrative review. EJNMMI Radiopharmacy and Chemistry.

Klassen et al. 2021. Chapter 1.1: History of the Sterile Insect Technique in Sterile Insect Technique: Principles And Practice In Area-Wide Integrated Pest Management. CRC Press. ISBN: 9780367474348.

Kleynhans et al. (2023). Fundamentals of Rhenium-188 Radiopharmaceutical Chemistry. Molecules.

Kluetz et al. 2014. Radium Ra 223 Dichloride Injection: U.S. Food and Drug Administration Drug Approval Summary. Clinical Cancer Research.

Knudsen. 1991. Legally-induced abortions in Denmark after Chernobyl. Biomedicine & Pharmacotherapy.

Kondev et al. 2021. The NUBASE2020 evaluation of nuclear physics properties. Chinese Physics C.

Kratochwil et al. 2016. 225Ac-PSMA-617 for PSMA-Targeted a-Radiation Therapy of Metastatic Castration-Resistant Prostate Cancer. Journal of Nuclear Medicine.

Kratz. 2021. Brush with Catastrophe: The Day the U.S. Almost Nuked Itself. Pieces of History, U.S. National Archives.

Kristensen et al. 2024. Estimated Global Nuclear Warhead Stockpiles. Federation of American Scientists.

L'Annunziata. 2016. Hall of Fame: Part III in Radioactivity: introduction and history, from the Quantum to Quarks. Elsevier. ISBN: 978-0-444-63489-4.

Laidra et al. 2017. Mental disorders among Chernobyl cleanup workers from Estonia: A clinical assessment. Psychological Trauma: Theory, Research, Practice, and Policy.

Lawrence Livermore National Laboratory. 2022a. LLNL's Fusion Ignition Shot Hailed as Historic Scientific Feat.

Lawrence Livermore National Laboratory. 2022b. Lawrence Livermore National Laboratory achieves fusion ignition. YouTube.

Lawrence Livermore National Laboratory. nd. Anatomy of a NIF Shot.

Lawrence. 1951. Nobel Lecture: The evolution of the cyclotron. The Nobel Prize.

Lazard. 2023. 2023 Levelized Cost of Energy.

Le et al. 2014. Methods of Increasing the Performance of Radionuclide Generators Used in Nuclear Medicine: Daughter Nuclide Build-Up Optimisation, Elution-Purification-Concentration Integration, and Effective Control of Radionuclidic Purity. Molecules.

Lee and Bairstow. 2015. Radioisotope Power Systems Reference Book for Mission Designers and Planners. Jet Propulsion Laboratory.

Lehtonen. 2023. The Governance Ecosystem of Radioactive Waste Management in France: Governing of and with Mistrust in The Future of Radioactive Waste Governance. ISBN: 978-3-658-40496-3.

Lelieveld et al. 2019. Effects of fossil fuel and total anthropogenic emission removal on public health and climate. Proceedings of the National Academy of Sciences.

Leung et al. 2019. Trends in Solid Tumor Incidence in Ukraine 30 Years After Chernobyl. Journal of Global Oncology.

Lindblom. 2018. Plutonium-238 Production for Space Exploration. American Chemical Society.

Lindén et al. 2021. 227Th-Labeled Anti-CD22 Antibody (BAY 1862864) in Relapsed/Refractory CD22-Positive Non-Hodgkin Lymphoma: A First-in-Human, Phase I Study. Cancer Biotherapy and Radiopharmaceuticals.

Lindquist et al. 1992. The New World screwworm fly in Libya: a review of its introduction and eradication. Medical and Veterinary Entomology.

Liu et al. 2017. A half-wave rectified alternating current electrochemical method for uranium extraction from seawater. Nature Energy.

Livingood and Seaborg. 1938. Radioactive Isotopes of Iodine. Physical Review.

Lovering et al. 2016. Historical construction costs of global nuclear power reactors. Energy Policy.

Lovering et al. 2022. Land-use intensity of electricity production and tomorrow's energy landscape. PLUS ONE.

Lu et al. 2009. Global potential for wind-generated electricity. Proceedings of the National Academy of Sciences.

Luchsinger et al al. 2021. Water within a permanently shadowed lunar crater: Further LCROSS modeling and analysis. Icarus.

Lundberg. 2019. Norway approves copper mine in Arctic described as 'most environmentally damaging project in country's history'.

Luo et al. 2022. Global and regional trends in incidence and mortality of female breast cancer and associated factors at national level in 2000 to 2019. Chinese Medical Journal.

Lyall. 2020. Project Mars.

MacDonald. 2022. Subsidies for Drax biomass. Ember.

Macrotrends. 2024. Crude Oil Prices - 70 Year Historical Chart.

Mahanti. 2007. Homi Jehangir Bhabha. Vigyan Prasar Science Portal.

Mahesh et al. 2023. Patient Exposure from Radiologic and Nuclear Medicine Procedures in the United States and Worldwide: 2009–2018. Radiology.

Malik. 1985. Yields of the Hiroshima and Nagasaki nuclear explosions (LA--8819, 1489669). Los Alamos National Laboratory.

Martins et al. 2012. Brazil nuts: determination of natural elements and aflatoxin. Acta Amazonica.

Mathieu and Rodés-Guirao. 2022. What are the sources for Our World in Data's population estimates? Our World in Data.

Matsuo. 2022. Re-Defining System LCOE: Costs and Values of Power Sources. Energies.

Mauro et al. 2005. Assessment of Variations in Radiation Exposure in the United States. U.S. Environmental Protection Agency.

Mayo Clinic Health System. 2023. Targeted radiation therapy: Y-90 and liver cancer.

Mayo Clinic. 2024a. Thyroid cancer: diagnosis and treatment.

Mayo Clinic. 2024b. Sodium Iodide I 131 (Oral Route).

Mayo. 2024. The largest emitters in the UK: annual review. Ember

Mbow et al. 2019. Chapter 5: Food Security in Climate Change and Land: an IPCC special report on climate change, desertification, land degradation, sustainable land management, food security, and greenhouse gas fluxes in terrestrial ecosystems. Intergovernmental Panel on Climate Change.

McLaughlin et al. 2000a. Chapter 2: Reactor and Critical Experiments Accidents in A Review of Criticality Accidents 2000 Revision. Los Alamos National Laboratory.

McLaughlin et al. 2000b. A Review of Criticality Accidents. Los Alamos National Laboratory.

McLean et al. 2017. A restatement of the natural science evidence base concerning the health effects of low-level ionizing radiation. Proceedings of the Royal Society B: Biological Sciences.

McMillan and Ableson. 1940. Radioactive Element 93. Physical Review.

McMillan. 1939. Radioactive Recoils from Uranium Activated by Neutrons. Physical Review.

McMillan. 1951. The Transuranium Elements: Early History. Nobel Lecture.

Meija et al. 2016. Isotopic compositions of the elements 2013 (IUPAC Technical Report). Pure and Applied Chemistry.

Meitner and Frisch. 1939. Disintegration of Uranium by Neutrons: a New Type of Nuclear Reaction. Nature.

Meitner et al. 1937. Über die Umwandlungsreihen des Urans, die durch Neutronenbestrahlung erzeugt warden. Zeitschrift für Physik.

Mettler et al. 2007. Health effects in those with acute radiation sickness from the Chernobyl accident. Health Physics.

Mettler et al. 2008. Effective Doses in Radiology and Diagnostic Nuclear Medicine: A Catalog. Radiology.

Mettler et al. 2020. Patient Exposure from Radiologic and Nuclear Medicine Procedures in the United States: Procedure Volume and Effective Dose for the Period 2006–2016. Radiology.

Meusburger et al. 2020. Plutonium aided reconstruction of caesium atmospheric fallout in European topsoils. Scientific Reports.

Miller. 2022. Video, photos capture first full-scale H-bomb test 70 years ago (LA-UR-22-31440). Los Alamos National Laboratory.

Mining Technology. 2019. Copper, fjords, reindeer and controversy: inside Norway's new arctic mine.

Ministry of Defence. 2020. Chapter 33: exposure to cosmic radiation (December 2020) in Management of radiation protection in defence: part 2 guidance (JSP 392).

Mitchell III and Turner. 1971. Breeder Reactors. US Atomic Energy Commission.

Mortazavi et al. 2019. Is Induction of Anomalies in Lymphocytes of the Residents of High Background Radiation Areas Associated with Increased Cancer Risk? Journal of Biomedical Physics and Engineering.

Müller et al. 2019. Terbium-161 for PSMA-targeted radionuclide therapy of prostate cancer. European Journal of Nuclear Medicine and Molecular Imaging.

Muller. 1946. The Production of Mutations. The Nobel Prize.

Murchie and Reid. 2020. Chapter 9: Uranium conversion and enrichment in Advances in Nuclear Fuel Chemistry. Elsevier.

Nair et al. 2009. Background radiation and cancer incidence in Kerala, India-Karanagappally cohort study. Health Physics.

NASA Earth Observatory. 2021. Ten Years After the Tsunami.

NASA. 1968. Saturn V Flight Manual.

NASA. 2021. Nuclear Propulsion Could Help Get Humans to Mars Faster, press release.

NASA. nd. Voyager.

Nath. 2022. Homi J Bhabha: a renaissance man among scientists. Niyogi Books. ISBN: 9391125115.

Nathwani et al. 2016. Polonium-210 poisoning: a first-hand account. The Lancet.

National Archives and Records Administration. nd. Nuclear Test Ban Treaty.

National Archives. 1985. Joint Soviet-United States Statement on the Summit Meeting in Geneva.

National Cancer Institute. 2020. Incidence Rate Report by State.

National Cancer Institute. 2021. Advanced Prostate Cancer, Radiopharmaceutical Improves Survival.

National Geographic. 2019. Why the government breeds and releases billions of flies a year. National Geographic.

National Geographic. 2021. Photos: A decade after disaster, wildlife abounds in Fukushima.

National Geographic. 2022. Can Norway balance its green energy goals with Indigenous concerns?

National Grid ESO. 2024. Historic generation mix and carbon intensity.

National Research Council U.S. 2014. Lessons learned from the Fukushima nuclear accident for improving safety of U.S. nuclear plants. ISBN: 978-0-309-27253-7.

NBC News. 2023. Exclusive: British PM Rishi Sunak says China is the 'biggest state threat' to economic interests. YouTube.

Neff. 1991. A Grand Uranium Bargain. New York Times.

Neidell *et al*. 2021. The unintended effects from halting nuclear power production: Evidence from Fukushima Daiichi accident. Journal of Health Economics.

Net Zero Tracker. 2024. Data explorer.

New York Times. 1987. Rise in retarded children predicted from Chernobyl.

New York Times. 2022. Scientists Achieve Nuclear Fusion Energy Breakthrough in the US.

Nobel Lectures. 1964. Chemistry 1942-1962. Elsevier Publishing Company.

Normile. 2016. Five years after the meltdown, is it safe to live near Fukushima? Science.

Nuclear Decommissioning Authority. 2008. Plutonium Options for Comment.

Nuclear Decommissioning Authority. 2023. UK Radioactive Waste and Material Inventory 2022.

Nuclear Energy Agency and the International Atomic Energy Agency. 2023. Uranium 2022: Resources, Production and Demand. OECD Publishing.

Nuclear Forensics International Technical Working Group. 2024. ITWG.

O'Donnell *et al.* 1981. Assessment of radiation doses from residential smoke detectors that contain americium-241. Oak Ridge National Laboratory.

Oak Ridge Associated Universities. nd. Why Did They Call It That?.

OECD and Nuclear Energy Agency. 2002. Chernobyl: Assessment of Radiological and Health Impacts: 2002 Update of Chernobyl: Ten Years On. ISBN: 978-92-64-18487-9.

OECD Nuclear Energy Agency and Internation Atomic Energy Agency. 2001. Management of depleted uranium.

OECD/NEA. 2021. Strategies and Considerations for the Back End of the Fuel Cycle.

Office for National Statistics. 2024. Gross Domestic Product: chained volume measures: Seasonally adjusted £m.

Ojovan *et al.* 2019. Performance of Wasteform Materials in An Introduction to Nuclear Waste Immobilisation. Elsevier. ISBN: 978-0-08-102702-8.

Orano. 2020. Rapport d'information du site de la Hague: Édition 2019.

Orano. 2022. Recycled uranium for low-carbon electricity.

Orano. 2023. Rapport d'information du site Orano la Hague: Édition 2022.

Orano. 2024. Rapport d'information du site Orano la Hague: Édition 2023.

Orano. nd,a. Orano la Hague.

Orano. nd,b. MOX, a fuel assembly made from recycled nuclear fuel.

Orano. nd,c. All about used fuel processing and recycling.

Organisation for Economic Co-operation and Development. 2022. Plastic pollution is growing relentlessly as waste management and recycling fall short, says OECD.

Our World in Data. 2024. Energy use per person vs. GDP per capita.

Owen. 2016. The Litvinenko inquiry: report into the death of Alexander Litvinenko. Home Office. ISBN: 978-1-4741-2734-9.

Ozasa *et al.* 2012. Studies of the Mortality of Atomic Bomb Survivors, Report 14, 1950–2003: An Overview of Cancer and Noncancer Diseases. Radiation Research.

Pallares *et al.* 2022. Development of radiopharmaceuticals for targeted alpha therapy: Where do we stand? Frontiers in Medicine.

Pavlov and Rybachenkov. 2013. Looking Back: The U.S.-Russian Uranium Deal: Results and Lessons. Arms Control Association.

Peace Research Institute Oslo. 2017. Battle deaths v3.1.

Peeva. 2021. Cyclotrons – What are They and Where Can you Find Them. International Atomic Energy Agency.

Pellow et al. 2015. Hydrogen or batteries for grid storage? A net energy analysis. Energy & Environmental Science.

Plompen et al. 2020. The joint evaluated fission and fusion nuclear data library, JEFF-3.3. The European Physical Journal A.

Ponomarev et al. 2022. Keystone and stumbling blocks in the use of ionizing radiation for recycling plastics. Radiation Physics and Chemistry.

Portugal-Pereira et al. 2018. Better late than never, but never late is better: Risk assessment of nuclear power construction projects. Energy Policy.

Posiva Oy. 2023. Posiva Oy animation of the Encapsulation plant. YouTube.

Posiva. 2023. 40 years today since Decision-in-Principle on final disposal of spent nuclear fuel in Finland.

Posiva. nd,a. Final disposal.

Radiant Energy Group. 2023. Public Attitudes toward Clean Energy: Nuclear Energy.

Radiation and Nuclear Safety Authority. 2015. Review of safety assessment in Posiva's construction license application for a repository at Olkiluoto.

Radiation Effects Research Foundation. 2015. Introduction to the Radiation Effects Research Foundation.

Radiation Effects Research Foundation. 2016. A Brief Description of the Radiation Effects Research Foundation.

Rae. 1978. Chapter 1: Selecting Heavy Water Processes in Separation of Hydrogen Isotopes.

Rahu et al. 2023. Suicide and other causes of death among Chernobyl cleanup workers from Estonia, 1986 – 2020: an update. European Journal of Epidemiology.

Ras et al. 2019. Carbon capture and utilization in the steel industry: challenges and opportunities for chemical engineering. Current Opinion in Chemical Engineering.

Rebello et al. 2021. Prostate cancer. Nature Reviews Disease Primers.

Reconstruction Agency. 2020. Number of earthquake-related deaths in the Great East Japan Earthquake (survey results as of September 30, 2020).

Reichard. nd. New world screwworm and its appearance in the eastern hemisphere. Food and Agriculture Organization of the United Nations.

Reichenberg et al. 2018. The marginal system LCOE of variable renewables – Evaluating high penetration levels of wind and solar in Europe. Energy.
Renew Economy. 2022. In Germany, a wind farm is dismantled to make way for expanded lignite coal mine.
Reuters. 2011. German government wants nuclear exit by 2022 at latest.
Reuters. 2017. Nineteen nations say they'll use more bioenergy to slow climate change.
Reuters. 2019. Norway gives go-ahead to disputed Arctic copper mine.
Reuters. 2022. Germany extends run times for coal-fired power plants to boost supply.
Reuters. 2023a. Spain confirms nuclear power phase-out, extends renewable projects deadlines.
Reuters. 2023b. What Macron means when he says he wants to 'take back control' of French power prices.
Reuters. 2023c. Nuclear life extension plans tested by obsolete components.
Reuters. 2023d. Fukushima: Why is Japan releasing water and is it safe?.
Reuters. 2023e. Putin revokes Russian ratification of global nuclear test ban treaty.
Reuters. 2024. Exclusive: Gulf state UAE considers a second nuclear power plant.
Reynolds et al. 2019. Contrapositive logic suggests space radiation not having a strong impact on mortality of US astronauts and Soviet and Russian cosmonauts. Scientific Reports.
Rhodes. 1986. Chapter 13: The New World in The Making of the Atomic Bomb. Simon & Schuster. ISBN: 978-0-671-44133-3.
Riesz. 1995. The life of Wilhelm Conrad Roentgen. American Journal of Roentgenology.
Ritchie and Rosado. 2022. Natural Disasters. Our World in Data.
Ritchie and Rosado. 2024. Energy mix. Our World in Data.
Ritchie et al. 2023. Internet. Our World in Data.
Ritchie et al. 2024a. Clean Water. Our World in Data.
Ritchie et al. 2024b. Access to Energy. Our World in Data.
Ritchie. 2022a. How does the land use of different electricity sources compare? Our World in Data.
Ritchie. 2022b. What are the safest and cleanest sources of energy? Our World in Data.

Ritchie. 2022c. The state of the world's rhino populations. Our World in Data.

Robertson *et al*. 2018. Development of 225Ac Radiopharmaceuticals: TRIUMF Perspectives and Experiences. Current Radiopharmaceuticals.

Rolls-Royce SMR. nd. Why Rolls-Royce SMR?.

Ronen. 2006. A Rule for Determining Fissile Isotopes. Nuclear Science and Engineering.

Röntgen. 1896. On a New Kind of Rays. Nature.

Rosenblat *et al*. 2010. Sequential Cytarabine and α-Particle Immunotherapy with Bismuth-213–Lintuzumab (HuM195) for Acute Myeloid Leukemia. Clinical Cancer Research.

Roser. 2020. The short history of global living conditions and why it matters that we know it. Our World in Data.

Roser. 2022. Malaria: One of the leading causes of child deaths, but progress is possible and you can contribute to it. Our World in Data.

Roser. 2023a. Extreme poverty: how far have we come, how far do we still have to go? Our World in Data.

Roser. 2023b. Mortality in the past: every second child died. Our World in Data.

Royal Society of Chemistry. nd. Uranium.

Ruiz De Porras *et al*. 2021. Chemotherapy in metastatic castration-resistant prostate cancer: Current scenario and future perspectives. Cancer Letters.

Rundo *et al*. 1977. Ingestion of 241Am Sources Intended for Domestic Smoke Detectors: Report of a Case. Health Physics.

Rutherford. 1900. A radio-active substance emitted from thorium compounds. The London, Edinburgh, and Dublin Philosophical Magazine and Journal of Science.

Rutherford. 1919. LIV. Collision of α particles with light atoms. IV. An anomalous effect in nitrogen. The London, Edinburgh, and Dublin Philosophical Magazine and Journal of Science.

Rutherford. 1927. Address of the President, Sir Ernest Rutherford, at the anniversary meeting, November 30, 1926. Proceedings of the Royal Society of London. Series A, Containing Papers of a Mathematical and Physical Character.

Sagan. 1983. The Day After. ABC News Viewpoint, broadcast on 20th November 1983.

Salonen et al. 2021. Status report of the Finnish spent fuel geologic repository programme and ongoing corrosion studies. Materials and Corrosion.

Sartor et al. 2014. Effect of radium-223 dichloride on symptomatic skeletal events in patients with castration-resistant prostate cancer and bone metastases: results from a phase 3, double-blind, randomised trial. The Lancet Oncology.

Sathekge et al. 2024. Actinium-225-PSMA radioligand therapy of metastatic castration-resistant prostate cancer (WARMTH Act): a multicentre, retrospective study. The Lancet Oncology.

Savanta and Radiant Energy Group. 2023. Public Attitudes toward Clean Energy: Nuclear Energy. Public Attitudes toward Clean Energy Index.

Schlesinger. 2018. Are wood pellets a green fuel? Science.

Schlömer et al. 2014. Annex III. Technology-specific Cost and Performance Parameters. Climate Change 2014: Mitigation of Climate Change. Contribution of Working Group III to the Fifth Assessment Report of the Intergovernmental Panel on Climate Change.

Schmidt. 1898. Ueber die von den Thorverbindungen und einigen anderen Substanzen ausgehende Strahlung. Annalen der Physik.

Schneider and Froggatt. 2019. World Nuclear Industry Status Report 2019. A Mycle Schneider Consulting.

Schneider and Froggatt. 2022. World Nuclear Industry Status Report 2022. A Mycle Schneider Consulting.

Schneider and Froggatt. 2023. World Nuclear Industry Status Report 2023. A Mycle Schneider Consulting.

Schneider and Froggatt. 2023. World Nuclear Industry Status Report 2023. Mycle Schneider Consulting.

Schneider et al. 2001. Systematic Radiological Assessment of Exemptions for Source and Byproduct Materials. U.S. Nuclear Regulatory Commission.

Seaborg. 1968. Need We Fear Our Nuclear Future? Bulletin of the Atomic Scientists.

Seaborg. 2019. The life and contributions to the periodic table of Glenn T. Seaborg, the first person to have an element named after him while he was still alive. Pure and Applied Chemistry.

Selby et al. 2021. A New Yield Assessment for the Trinity Nuclear Test, 75 Years Later. Nuclear Technology.

Sellafield Ltd. 2023. Sellafield Ltd Annual Review of Environmental Performance 2022/23.

Shea. 2018. The Journey of Actinium-225: How Scientists Discovered a New Way to Produce a Rare Medical Radioisotope. U. S. Department of Energy.

Shepherd. 1990. Historical development of the windmill. NASA Contractor Report 4337. DOI: 10.2172/6342767.

Siegbahn. 1939. The Nobel Prize in Physics 1939: Award ceremony speech. The Nobel Prize.

Sierra Club. 2016. Nuclear Power.

Siwowska et al. 2019. Therapeutic Potential of 47Sc in Comparison to 177Lu and 90Y: Preclinical Investigations. Pharmaceutics.

Skjoeldebrand. 1973. The Need for Fast Breeder Reactors. Bulletin, International Atomic Energy Agency.

Sky News. 2023. Cold snap forces National Grid to put three coal power generators on standby.

Smil. 2016 Energy Transitions: Global and National Perspectives (Second expanded and updated edition). Praeger. ISBN: 144085324X.

Smith et al. 2023. The risks of radioactive wastewater release. Science.

Soddy. 1920. The interpretation of radium, and the structure of the atom. G. P. Putnam's Sons.

Sovacool et al. 2016. Balancing safety with sustainability: assessing the risk of accidents for modern low-carbon energy systems. Journal of Cleaner Production.

Space Policy and Politics. 2017. Elon Musk, National Governors Association, July 15, 2017. YouTube.

Spalding et al. 2005. Age written in teeth by nuclear tests. Nature.

Spinelli and Osborn. 1991. The effects of the Chernobyl explosion on induced abortion in Italy. Biomedicine & Pharmacotherapy.

SSE. 2023. Scotland's largest offshore wind farm now fully operational.

Stamoulis et al. 1999. Strontium-90 concentration measurements in human bones and teeth in Greece. Science of The Total Environment.

Sterman et al. 2018. Does replacing coal with wood lower CO_2 emissions? Dynamic lifecycle analysis of wood bioenergy.

Stoner et al. 2021. Household cooking fuel estimates at global and country level for 1990 to 2030. Nature Communications.

Stop Sizewell C. 2021. Sizewell C and Climate Change.
Stopar. nd. Exploring the Moon's South Pole. Lunar and Planetary Institute.
Suh. 2023. The Strategic Nuclear Forces Dataset v2.0.
Sutou. 2016. A message to Fukushima: nothing to fear but fear itself. Genes and Environment.
Sveriges Radio. 2019. 25 years after Chernobyl, how Sweden found out.
Tammemagi and Jackson. 2009. Chapter 2: Splitting the Atom in Half-lives: a guide to nuclear technology in Canada. Oxford University Press. ISBN: 978-0-19-543152-0.
Taz. 2019. Thunbergs Segelreise in die USA: Gretas Törn schädlicher als Flug.
Tesla. 2024. European Union Energy Label.
The Atlantic. 2022. Is This the 'Kitty Hawk Moment' for Fusion Energy?.
The Danish Environmental Protection Agency. 2018. Life Cycle Assessment of grocery carrier bags.
The Golden Goose Award. 2016. 2016: The Sex Life of the Screwworm Fly.
The Guardian. 2019. Greta Thunberg sets sail for New York on zero-carbon yacht.
The Japan Times. 2023. Over 20 nations, including Japan, call for tripling of nuclear energy.
The Planning Inspectorate. 2021. The Sizewell C Project, Environmental Statement Addendum.
The Royal Society. 2020. Nuclear cogeneration: civil nuclear energy in a low-carbon future policy briefing.
The Times. 2019. UK skipper flew in for Greta Thunberg's Atlantic crossing.
The Washington Post. 2019. Swedish climate activist Greta Thunberg is sailing to America amid a storm of online attacks.
The White House. 2022. Joint Statement of the Leaders of the Five Nuclear-Weapon States on Preventing Nuclear War and Avoiding Arms Races.
The Whitehaven News. 2006. October 17 1956 Her Majesty the Queen opens Calder Hall.
The World Bank. 2020. Minerals for Climate Action: The Mineral Intensity of the Clean Energy Transition.
The World Bank. 2023. Access to electricity (% of population).
The World Bank. 2024a. GDP per capita (constant 2015 US$).
The World Bank. 2024b. Population, total.

Tietze et al. 2023. Topical ^{188}Re Ionizing Radiation Therapy Exerts High Efficacy in Curing Nonmelanoma Skin Cancer. Clinical Nuclear Medicine.

Time. 1950. Foreign News: The Explosion and All.

Toon et al. 2019. Rapidly expanding nuclear arsenals in Pakistan and India portend regional and global catastrophe. Science Advances.

Tran et al. 2006. Rice Mutation Breeding in Institute of Agricultural Genetics, Viet Nam. International Atomic Energy Agency: Plant Mutation Reports.

Trauth et al. 1993. Expert Judgment on Markers to Deter Inadvertent Human Intrusion into the Waste Isolation Pilot Plant. United States Department of Energy.

Trichopoulos et al. 1987. The victims of Chernobyl in Greece: induced abortions after the accident. BMJ.

Tsouris et al. 2017. Nature Energy.

TVO. nd. OL3.

U. S. Department of Energy. nd. Einstein's Letter.

U. S. Food and Drug Administration. 2022. FDA approves Pluvicto for metastatic castration-resistant prostate cancer.

U.S. Department of Energy. 2013. Inside the 60-Inch Cyclotron (January 18, 1939). Flickr.

U.S. Department of Energy. 2017. Isotopes Routinely Available: radium-223 and thorium-227.

U.S. Department of Energy. nd. Bomb Design and Components. The Manhattan Project: An Interactive History.

U.S. Department of Energy. 2024. U.S. Sets Targets to Triple Nuclear Energy Capacity by 2050.

U.S. Department of State. 2022. The Nuclear Forensics International Technical Working Group 2022 Annual Meeting.

U.S. Department of State. nd. CTBT: Annex 2 States.

U.S. Food & Drug Administration. 2016. Food Irradiation: What You Need to Know.

U.S. Geological Survey. 2024. Earthquake Catalog.

U.S. National Library of Medicine. 2023a. Venetoclax and Lintuzumab-Ac225 in AML Patients, NCT03867682.

U.S. National Library of Medicine. 2023b. Targeted Alpha-emitter Therapy of PRRT Naïve and Previous PRRT Neuroendocrine Tumor Patients (ALPHAMEDIX02), NCT05153772.

U.S. National Library of Medicine. 2023c. First-in-human Study of BAY2287411 Injection, a Thorium-227 Labeled Antibody-chelator Conjugate, in Patients With Tumors Known to Express Mesothelin, NCT03507452.

U.S. National Library of Medicine. 2024a. Actinium 225 Labeled Anti-CEA Antibody (Ac225-DOTA-M5A) for the Treatment of CEA Producing Advanced or Metastatic Cancers, NCT05204147.

U.S. National Library of Medicine. 2024b. Targeted Alpha Particle Radiotherapy for Metastatic Uveal Melanoma, NCT05496686.

U.S. National Library of Medicine. 2024c. Radioimmunotherapy (111Indium/225Actinium-DOTA-daratumumab) for the Treatment of Relapsed/Refractory Multiple Myeloma, NCT05363111.

U.S. National Library of Medicine. 2024d. Targeted Alpha Therapy Using Astatine (At-211) Against Differentiated Thyroid Cancer, NCT05275946.

U.S. National Library of Medicine. 2024e. A Safety Study of 212Pb-Pentixather Radioligand Therapy, NCT05557708.

UChicago News. nd. The first nuclear reactor, explained.

UK Government. 2020. The Ten Point Plan for a Green Industrial Revolution.

UK Health Security Agency. 2023. Ionising Radiation and you.

UK Health Security Agency. nd. Ionising radiation and you.

UK Research and Innovation. 2021. A brief history of climate change discoveries.

UK Space Agency. 2024. £33 million boost for national space programme.

UN Treaty Collection. 2024. Comprehensive Nuclear-Test-Ban Treaty.

UN. Secretary-General. 1987. Report of the World Commission on Environment and Development.

United Nations Climate Change. 2018. Transcript of People's Seat Address by Sir David Attenborough at COP24.

United Nations Climate Change. 2022. António Guterres: Time to Jump-Start the Renewable Energy Transition.

United Nations Development Programme. 2024. Human Development Index (HDI).

United Nations Economic Commission for Europe. 2022. Carbon Neutrality in the UNECE Region: Integrated Life-cycle Assessment of Electricity Sources.

United Nations Office for Disarmament Affairs. 2024a. Treaties Database.

United Nations Office for Disarmament Affairs. 2024b. Treaty on the Non-Proliferation of Nuclear Weapons.

United Nations Scientific Committee on the Effects of Atomic Radiation. 2008a. Sources and Effects of Ionizing Radiation, Volume I: Sources. United Nations. ISBN: 978-92-1-142274-0.

United Nations Scientific Committee on the Effects of Atomic Radiation. 2008b. Sources and Effects of Ionizing Radiation, Volume II: Effects. United Nations. ISBN: 978-92-1-142280-1.

United Nations Scientific Committee on the Effects of Atomic Radiation. 2013. UNSCEAR 2013 Report Volume I. United Nations. ISBN: 978-92-1-142291-7.

United Nations Scientific Committee on the Effects of Atomic Radiation. 2017. Sources, effects and risks of ionizing radiation, Annex C: Biological effects of selected internal emitters-Tritium. United Nations. ISBN: 978-92-1-142316-7.

United Nations Scientific Committee on the Effects of Atomic Radiation. 2018. Evaluation of data on thyroid cancer in regions affected by the Chernobyl accident. United Nations.

United Nations Scientific Committee on the Effects of Atomic Radiation. 2022a. UNSCEAR 2020/2021 Report Volume II. United Nations. ISBN: 978-92-1-139207-4.

United Nations Scientific Committee on the Effects of Atomic Radiation. 2022b. Sources, effects and risks of ionizing radiation: UNSCEAR 2020/2021 report to the General Assembly, with scientific annexes. ISBN: 978-92-1-139206-7.

United Nations Treaty Collection. nd. Partial Nuclear-Test-Ban Treaty (PTBT).

United Nations. 1968. Treaty on the Non-Proliferation of Nuclear Weapons.

United Nations. 2018. UN chief: Climate change poses 'existential threat' to humanity. YouTube.

United Nations. 2022a. Renewables: Cheapest form of power.

United Nations. 2022b. Recounting Horrors in Hiroshima, Nagasaki, Atomic Bombing Survivors Urge Recommitment to 'Never Again' Use Nuclear Weapons, as Review Conference Wraps Up First Week.

United Nations. 2023. António Guterres (UN Secretary-General) at the Opening ceremony of the World Climate Action Summit | COP28, UN Climate Change Conference. UN Wev TV.

United Nations. 2024a. Cancer rates set to rise 77 per cent by 2050.

United Nations. 2024b. Constitution of the World Health Organization.

United Nations. 2024c. Charter of the United Nations And Statute of the International Court of Justice.

United States Department of Agriculture. 2018. New world screwworm: what you need to know. Animal and Plant Health Inspection Service.

United States Department of Energy. 2002. Restricted data declassification decisions 1946 to the present (RDD-8).

United States Nuclear Regulatory Commission. 2019. Backgrounder on Tritium, Radiation Protection Limits, and Drinking Water Standards.

United States Nuclear Regulatory Commission. 2021. Units of radiation dose.

Unnerstall. 2017. How expensive is an energy transition? A lesson from the German Energiewende. Energy, Sustainability and Society.

Uppsala Conflict Data Program. 2023. Georeferenced Event Dataset v23.1.

Uri. 2019. 40 Years Ago: Voyager 1 Explores Jupiter. NASA.

Uri. 2020. 40 Years Ago: Voyager 1 Explores Saturn. NASA.

US Atomic Energy Commission. 1944. Notes On Meeting of April 26, 1944.

US Atomic Energy Commission. 1958. Peaceful uses of fusion.

US Department of Energy. 2010. Ionizing Radiation Dose Ranges.

US Department of Energy. 2021. Energy For Space.

US Department of Energy. 2023. At COP28, Countries Launch Declaration to Triple Nuclear Energy Capacity by 2050, Recognizing the Key Role of Nuclear Energy in Reaching Net Zero.

US Deptartment of Labor. 2022. List of Goods Produced by Child Labor or Forced Labor.

US Government. 1993. Agreement between the Government of the United States of America and the Government of the Russian Federation Concerning the Disposition of Highly Enriched Uranium Extracted from Nuclear Weapons.

Vásquez et al. 2023. The Human Freedom Index 2023: A Global Measurement of Personal, Civil, and Economic Freedom. The Cato Institute.

Veritasium. 2016. Chernobyl - What It's Like Today. YouTube.

Viollet. 2017. From the water wheel to turbines and hydroelectricity. Technological evolution and revolutions. Comptes Rendus Mecanique. 2017. DOI: 10.1016/j.crme.2017.05.016.

Vogel et al. 2021. Challenges and future options for the production of lutetium-177. European Journal of Nuclear Medicine and Molecular Imaging.

von Baeckmann et al. 1995. Nuclear verification in South Africa.

Voss. 1984. SNAP Reactor Overview. Defense Technical Information Center.

Wadia. 2015. Excerpts from the Speech on Occasion of the Inauguration of ICTS Campus.

Wallenius et al. 2020. Nuclear Forensics at Jrc – From Ad-Hoc Analysis To Full-Grown Discipline. The Nuclear Forensics International Technical Working Group (ITWG): A Unique Forum of Practitioners Contributing to Global Nuclear Security.

Wang et al. 2022. Prostate Cancer Incidence and Mortality: Global Status and Temporal Trends in 89 Countries From 2000 to 2019. Frontiers in Public Health.

Wang et al. 2024. Updated Mining Footprints and Raw Material Needs for Clean Energy. The Breakthrough Institute.

Webb. 1949. The Fogging of Photographic Film by Radioactive Contaminants in Cardboard Packaging Materials. Physical Review.

Weinstock and Crist. 1948. The Vapor Pressure of Uranium Hexafluoride. The Journal of Chemical Physics.

Woddi et al. 2009. A Brief History of the Indian Nuclear Program in India's Nuclear Fuel Cycle. Springer. ISBN: 978-3-031-01361-4.

World Bank Group. 2020. Global Photovoltaic Power Potential by Country.

World Bank. 2023. GDP (current US$).

World Food Prize Foundation. nd. 1992: Knipling and Bushland.

World Health Organization. 2019. Mosquito sterilization offers new opportunity to control chikungunya, dengue, and Zika.

World Health Organization. 2020. A framework for mental health and psychosocial support in radiological and nuclear emergencies.

World Health Organization. 2022. Cancer.

World Health Organization. 2022. Guidelines for drinking-water quality.

World Health Organization. 2023. HIV and AIDS.

World Health Organization. 2024. WHO COVID-19 dashboard.
World Nuclear Association. 2017. Nuclear Power Economics and Project Structuring – 2017 Edition.
World Nuclear Association. 2020a. Heat Values of Various Fuels.
World Nuclear Association. 2020b. Cooling Power Plants.
World Nuclear Association. 2020c. National Policies and Funding.
World Nuclear Association. 2021a. Fast Neutron Reactors.
World Nuclear Association. 2021b. Research Reactors.
World Nuclear Association. 2022a. Uranium Enrichment.
World Nuclear Association. 2022b. Conversion and Deconversion.
World Nuclear Association. 2022c. Radioactive Waste Management.
World Nuclear Association. 2022d. RBMK Reactors – Appendix to Nuclear Power Reactors.
World Nuclear Association. 2022e. Chernobyl Accident 1986.
World Nuclear Association. 2023a. Nuclear Power in Belgium.
World Nuclear Association. 2023b. Plutonium.
World Nuclear Association. 2024a. Nuclear Energy in Denmark.
World Nuclear Association. 2024b. Nuclear Power Economics and Project Structuring: 2024 Edition.
World Nuclear Association. 2024c. Nuclear Power in Finland.
World Nuclear Association. 2024d. Storage and Disposal of Radioactive Waste.
World Nuclear Association. 2024e. Fukushima Daiichi Accident.
World Nuclear Association. nd. How is uranium made into nuclear fuel?
World Nuclear News. 2019. UK confirms funding for Rolls-Royce SMR.
World Nuclear News. 2022a. Macron sets out plan for French nuclear renaissance.
World Nuclear News. 2022b. Belgium government allocates funding for SMR research.
World Nuclear News. 2022c. China's demonstration HTR-PM reaches full power.
World Nuclear News. 2022d. NNL breakthrough to boost medical radio-isotope supply.
World Nuclear News. 2022e. Canadian Candu produces cancer therapy isotope.

World Nuclear News. 2023a. Sweden plans 'massive' expansion of nuclear energy.

World Nuclear News. 2023b. Swedish nuclear: Government moves to change law.

World Nuclear News. 2024a. EDF announces Hinkley Point C delay and rise in project cost.

World Nuclear News. 2024b. Chinese industrial nuclear steam project commissioned.

World Nuclear News. 2024c. Fuel loading begins at Indian fast breeder reactor.

World Nuclear News. 2024d. TerraPower breaks ground for Natrium plant.

World Wide Fund for Nature. 2021. Five Reasons to Oppose The Inclusion of Gas and Nuclear Power in the EU Taxonomy.

WWF. nd,a. Black Rhino.

WWF. nd,b. Javan Rhino.

WWF. nd,c. Rhino.

WWF. nd,d. African Rhinos.

Yadav and Banerjee. 2020. Net energy and carbon footprint analysis of solar hydrogen production from the high-temperature electrolysis process. Applied Energy.

Yang *et al.* 2017. Typhoon Nina and the August 1975 Flood over Central China. Journal of Hydrometeorology.

Yarris. 2021. The cyclotron's history at Berkeley. Berkeley College of Chemistry.

Zeitlin *et al.* 2013. Measurements of Energetic Particle Radiation in Transit to Mars on the Mars Science Laboratory. Science.

Zhang *et al.* 2022. Incidence and prognosis of thyroid cancer in children: based on the SEER database. Pediatric Surgery International.

Notes

Note to Introduction

1 Survey: Radiant Energy Group (2023). Greenpeace: Greenpeace (nd). Friends of the Earth: Friends of the Earth (2024). WWF: World Wide Fund for Nature (2021). 350.org: 350.org (2023). Sierra Club: Sierra Club (2016).

Chapter 1. Wresting Fire from the Gods

1 Becquerel (1896a).
2 Becquerel (1896b).
3 Schmidt: Schmidt (1898). Curie: Curie (1898).
4 Holden *et al.* (2018).
5 Rutherford (1919).
6 Blackett (1925).
7 Fermi (1934).
8 Hahn and Strassmann (1939).
9 Atomic Heritage Foundation (2022a).
10 Meitner and Frisch (1939).
11 American Institute of Physics (1967).
12 There are 0.00000000000000000016 – or 1.6×10^{-19} – joules in an electron-volt.
13 My calculation assumes a 10-watt lightbulb, an energy-density of 7.8 million kilowatt-hours per kilogram for uranium (when burned in a fast reactor, which we'll get to in Chapter 7) and 6.7 kilowatt-hours per kilogram for coal (World Nuclear Association, 2020a), and a thermal energy-to-electrical energy efficiency of 33 per cent (World Nuclear Association, 2020b).
14 Atomic Heritage Foundation (2022b).
15 U. S. Department of Energy (nd).
16 Bretscher and Cockcroft (1955).

17 Neutron energies: Duderstadt and Hamilton (1976a).
18 Rhodes (1986).
19 Fermi (1952).
20 Oak Ridge Associated Universities (nd).
21 Anderson (1966).
22 UChicago News (nd).
23 Fermi (1952).
24 American Institute of Physics (2007).

Chapter 2. How Nuclear Power Works

1 Gowlett (2016).
2 Waterwheels: Viollet 2017. Windmills: Shepherd 1990.
3 Ritchie and Rosado (2024).
4 Energy Institute (2024) and Smil (2016), with major processing by Ritchie and Rosado (2024).
5 Quality of life: Roser (2020). Poverty: Roser (2023a). Lifespan: Dattani et al. (2023a). Child mortality: Dattani et al. (2023b). Drinking water: Ritchie et al. (2024a). Internet: Ritchie et al. (2023). Wealth: The World Bank (2024a).
6 Data from U.S. Energy Information Administration (2023), Energy Institute (2024), and Mathieu and Rodés-Guirao (2022), with major processing by Our World in Data (2024).
7 World population: The World Bank (2024b). Electricity access: The World Bank (2023). Our World in Data: Ritchie et al. (2024b).
8 Long-term annual data: Energy Institute (2024). Sources today: Ember (2024).
9 Total electricity: Energy Institute (2024).
10 Idaho National Laboratory (2019).
11 Eisenhower's speech: International Atomic Energy Agency (nd).
12 Members: International Atomic Energy Agency (2023a).
13 Statute: International Atomic Energy Agency (1989).
14 Account of the day: The Whitehaven News (2006). As an aside, The Whitehaven News is my local paper.
15 International Atomic Energy Agency (2024a).
16 Breeze (2017).
17 World Nuclear Association (nd,a).
18 Reactor types: International Atomic Energy Agency (2024a). Electricity data: Ember (2024).

19 International Atomic Energy Agency (2024a).
20 European Nuclear Society (nd).
21 Breeze (2017).
22 Hagemann *et al.* (1970).
23 Holden *et al.* (2018).
24 Centrifuges: Murchie and Reid (2020). Boiling point: Royal Society of Chemistry (nd). Hex: Weinstock and Crist (1948).
25 World Nuclear Association (2022a).
26 '... 0.1 per cent ...': World Nuclear Association (2022a). Stockpile: World Nuclear Association (2022b).
27 Hydrogen-1 abundance: Hagemann *et al.* (1970). Heavy-water: Rae (1978).
28 A history of CANDU reactors: Tammemagi and Jackson (2009). Export: International Atomic Energy Agency (2024a).
29 A Tesla Model S does 3.6 miles to the kilowatt-hour (Tesla, 2024).
30 My calculation assumes an energy-density of 6.7 kilowatt-hours per kilogram for coal and 1.1 million kilowatt-hours per kilogram for 3.5 per cent enriched uranium in a light-water reactor (World Nuclear Association, 2020a), and a thermal energy-to-electrical energy efficiency of 33 per cent (World Nuclear Association, 2020b).
31 Ember (2024).
32 Human Development Index: United Nations Development Programme (2024). Human Freedom Index: Vásquez *et al.* (2022).
33 Ember (2024).
34 International Atomic Energy Agency (2024a).
35 My calculation uses the average light-water reactor power on a per-decade basis (International Atomic Energy Agency, 2024a) and assumes a capacity factor (see note 3, Chapter 4) the same as the 2019 global median (86 per cent; International Atomic Energy Agency, 2020a).
36 '... a million people.': my calculation uses the median capacity of reactors built since 2018 (1.1 gigawatts; International Atomic Energy Agency, 2024a) and a capacity factor of 86 per cent (see note 3, Chapter 4). '... 80 kilograms ...': assuming a reactor uses about 30 tonnes of fuel per year. This varies from reactor to reactor, but it's a good rule of thumb (World Nuclear Association, nd,a).
37 International Atomic Energy Agency (2024a)
38 International Atomic Energy Agency (2024a)
39 Global fleet: International Atomic Energy Agency (2024a). Electricity share: Ember (2024)

40 Ember (2024).
41 Data used to plot graph: International Atomic Energy Agency (2024a).
42 Schneider and Froggatt (2022).
43 Ember (2024).
44 Schneider and Froggatt (2023).
45 International Atomic Energy Agency (2023b).
46 German-led coalition: Euractiv (2023a). Hydrogen footprints: Yadav and Banerjee (2020). Austria: Euronews (2023a). Scotland: BBC News (2022a). Denmark: World Nuclear Association (2024a). Australia: Australian Government (2023). Spain: Reuters (2023a).

Chapter 3. Unreliables

1 HM Government (2019).
2 European Climate Law: European Parliament (2021). 'Other nations', Germany, Sweden, and 'net zero pledges: Net Zero Tracker (2024). British Antarctic Survey: British Antarctic Survey (2018).
3 Friedlingstein *et al.* (2023a).
4 World energy: Energy Institute (2024). Emissions: Friedlingstein *et al.* (2023a).
5 Guterres 2018: United Nations (2018). Guterres 2023: United Nations (2023). Attenborough: United Nations Climate Change (2018). Biden: CNBC (2020).
6 Electricity: Ember (2024). Energy: Energy Institute (2024).
7 International Energy Agency (nd,a).
8 Electricity: Ember (2024). Energy and emissions: International Energy Agency (nd,b).
9 International Energy Agency (nd,c).
10 International Energy Agency (nd,d).
11 Calculating CO_2 emissions by sector is complicated. Breakdowns depends on how you slice and dice the data. For example, it's easy to double-count emissions from electricity because we use it extensively in buildings and industry (but not so much for transport). For an overall breakdown of emissions by sector, I used sector data from the International Energy Agency (2023a) and total emissions from Friedlingstein *et al.* (2023a). Breakdowns from other sources attribute emissions slightly differently and thus give slightly different answers. Whichever way you cut it, though, the overall picture stays the same: decarbonising society is going to be difficult and is only possibly by transforming our energy systems.

12 International Energy Agency (nd,e).
13 International Energy Agency (nd,f).
14 De Ras et al. (2019).
15 International Energy Agency (2021).
16 Summary of how ammonia changed the world: Erisman et al. (2008). Human population: The World Bank (2024b). Food production and ammonia-based fertiliser use: Mbow et al. (2019).
17 International Energy Agency (2023b).
18 International Energy Agency (2023c).
19 Gas and electricity prices vary from place to place. I calculated these numbers by dividing the median price of electricity (per kilowatt-hour) by the median price of gas (per kilowatt-hour) in the European Union (plus the United Kingdom). The data span from 2008 to 2023. Taxes are included in my calculation, but even when they're excluded the overall picture is the same: electricity is still pricier than gas (although the range shrinks to 2.0–3.1 times more expensive). Electricity prices: Department for Energy Security and Net Zero (2024a). Gas prices: Department for Energy Security and Net Zero (2024b).
20 Lu et al. (2009).
21 Ten Point Plan: UK Government (2020). 'Evangelist': BBC News (2020a). 'Saudi Arabia': BBC News (2020b).
22 Calculated by dividing 30 gigawatts over a year (265,000 gigawatt-hours) by the UK's electricity consumption per person (4.7 megawatt-hours). Data: Ember (2024).
23 Ember (2024).
24 Median annual capacity factors based on 15 years of data (Ember 2024).
25 Calculated using wind capacity and generation data from Ember (2024).
26 Breakdown of the UK's electricity grid going back to 1 January 2009, half-hour by half-hour: National Grid ESO (2024).
27 Precautionary measures: Energy Voice (2022) and Sky News (2023). March 2023: BBC News (2023a). Heatwave: BBC News (2023b).
28 Ratcliffe-on-Soar: Independent (2024). UK's grid: Ember (2024).
29 Data used to plot graph: National Grid ESO (2024). I plotted a weekly moving average to smooth out the short-term noise.
30 Ember (2024).
31 Ember (2024).
32 Shell's (now-deleted) tweet: '@Shell_NatGas'. 'Smart partner': @bp_plc (2019a). 'Perfect partner': @bp_plc (2019b).

33 Musk's interview: Space Policy and Politics (2017).
34 Solar league table: World Bank Group (2020).
35 Electricity: Ember (2024). Energy: Energy Institute (2024).
36 Solar: Ember (2024). Politicking: Department for Business, Energy, and Industrial Strategy (2021).
37 International Energy Agency (2022a).
38 NBC News (2023).
39 Geth et al. (2015).
40 Calculated by dividing pumped hydro storage capacity (Geth et al. 2015) by electricity consumption on a country-by-country basis (Ember 2024). For 'Europe' I used EU data plus Norway, Switzerland, and the UK.
41 According to Tesla's website, a 3.9-megawatt-hour Megapack cost $1.04 million (or £810,000) in August 2024. The UK's annual electricity consumption per head is 4.7 megawatt-hours (Ember, 2024); 67,000 people, therefore, consume 316,000 megawatt-hours of electricity every year; dividing Cottingham's storage capacity (196 megawatt-hours; BBC News, 2022b) by 316,000 megawatt-hours and converting the number from years to hours yields 5 hours and 26 minutes.
42 The UK's total electricity consumption is 867,000 megawatt-hours per day (Ember, 2024); Cottingham can store 196 megawatt-hours of electricity (BBC News, 2022b); dividing the former by the latter yields 4,400 Cottingham-sized facilities. A Tesla Megapack costs £800,000 for 3.9 megawatt-hours, which equals £208,000 per megawatt-hour; 867,000 megawatt-hours divided by £208,000 per megawatt-hour yields £180 billion. For comparison, the UK's GDP in 2022 and 2023 was £2.3 trillion (Office for National Statistics, 2024).
43 Pellow et al. (2015).

Chapter 4. Net Zero Is Impossible Without Nuclear Power

1 A history of climate change: UK Research and Innovation (2021).
2 'Greatest climate act': TVO (nd). Europe's nuclear reactors: International Atomic Energy Agency (2024a).
3 Nuclear's median capacity factor is around 86 per cent (that's the number for 2019, but it doesn't change much from year to year; International Atomic Energy Agency, 2020a). I used data from Ember (2024) to calculate the global capacity factor of all sources; nuclear came out on top every year since 2000 (as far back as the Ember data go). A 1.6-gigawatt

reactor with a capacity factor of 86 per cent generates 12,000 gigawatt-hours of electricity per year; dividing this by average OECD electricity use (8 megawatt-hours per person; Ember, 2024) yields 1.5 million people.

4 Calculated using the UK's current energy demand (317,000 gigawatt-hours), solar generation (13,500 gigawatt-hours), wind generation (82,500 gigawatt-hours), hydro generation (5,200 gigawatt-hours), and fossil fuel generation (117,000 gigawatt-hours), and assuming an Olkiluoto-3-like reactor generates 12,000 gigawatt-hours annually (see previous note). Electricity data from Ember (2024).

5 Calculated by dividing Europe's annual fossil fuel electricity consumption (2,060,000 gigawatt-hours; Ember, 2024) by 12,000 gigawatt-hours (see footnote 3).

6 Europe's annual electricity demand is 4,950,000 gigawatt-hours (Ember, 2024); that amount *again* – plus the 2,060,000 gigawatt-hours it already consumes from fossil fuels – adds up to 7,010,000 gigawatt-hours; dividing that number by 12,000 gigawatt-hours (see footnote 3) gives ~ 580 reactors.

7 Between 1960 and 1990, Europe (excluding Russia) connected 213 nuclear reactors to the grid (International Atomic Energy Agency, 2024a).

8 The world's annual fossil fuel electricity consumption is 17,900,000 gigawatt-hours (Ember, 2024). Dividing that by 12,000 gigawatt-hours (see footnote 3) gives ~ 1,500 reactors.

9 A 160 per cent increase in the world's electricity demand (currently 29,500,000 gigawatt-hours; Ember, 2024) is an extra 47,200,000 gigawatt-hours. Dividing that by 12,000 gigawatt-hours (see footnote 3) gives ~ 3,900 reactors. Adding the ~ 1,500 reactors needed to replace current fossil fuel generation totals ~ 5,400 reactors.

10 Germany's deadlines: International Energy Agency (nd,g). Birol's quote: Clean Energy Wire (2017).

11 Unnerstall (2017) estimated *Energiewende* cost approximately €505–530 billion by 2025, and that total costs will be €600–700 billion. This estimate is consistent with the more recent estimate by Emblemsvåg (2024).

12 Electricity: Ember (2024). Energy: Energy Institute (2024). Quote: Der Spiegel (2019).

13 Wind turbine longevity: EDF (2019). Solar panel: EDF (2020). To illustrate nuclear power's longevity, consider this: the 58 reactors the USA connected to its grid in the 1970s have a median age of 50 years, and 70 per

cent of them are still working (International Atomic Energy Agency, 2024a).
14 Germany's electricity mix: Ember (2024). Merkel: Reuters (2011).
15 Germany's electricity mix: Ember (2024). Avoidable CO_2: Kharecha and Sato (2019). Emissions: Friedlingstein *et al.* (2023a).
16 Jarvis *et al.* (2022) estimated the social cost of *Energiewende* and attribute most of it to air pollution. They also estimated 800 excess deaths per year on average due to the nuclear phase-out between 2012 and 2019, which totals 5,600 excess deaths.
17 *Atomausstieg* success: BBC News (2023c). Street parties: Euronews (2023b).
18 Germany's electricity mix: Ember (2024).
19 Late-2022: Reuters (2022). Autumn 2023: Bloomberg (2023a). Electricity: Ember (2024).
20 Gabriel and Löfven's exchange: Financial Times (2014).
21 Renew Economy (2022).
22 CO_2 and electricity: Ember (2024). Swedish nuclear expansion: World Nuclear News (2023a). Pourmokhtari's quote: World Nuclear News (2023b).
23 Olkiluoto-3 cost: Schneider and Froggatt (2023). Assuming *Energiewende* cost €500 billion so far – which is on the cheap end of the estimate by Unnerstall (2017) – then Germany could have bought 40 (€500 billion ÷ €12.4 billion) Olkiluoto-3 reactors. Assuming an annual electricity generation of 12 terawatt-hours (see footnote 3), they'd net 480 terawatt-hours of nuclear electricity annually.
24 Those 480 terawatt-hours of nuclear electricity (see previous footnote) plus the 170 terawatt-hours Germany switched off since 2000 (Ember, 2024) would have totalled 650 terawatt-hours. Eliminating fossil fuels (234 terawatt-hours), wind (137 terawatt-hours), and solar (62 terawatt-hours) leaves 217 terawatt-hours (Ember, 2024). Collectively, Germany drives its 48.5 million cars (Eurostat, 2024a) a combined distance of 743 billion kilometres annually (Federal Ministry of Digital and Transport, 2023) which, assuming an electric car drives 5.6 kilometres per kilowatt-hour (EDF, nd), would require 133 terawatt-hours if they were all electric; that leaves 84 terawatt-hours. Assuming it takes 51 kilowatt-hours of electricity to make a kilogram of green hydrogen (International Renewable Energy Agency, 2020), Germany could make 1.7 million tonnes of green hydrogen per year. The global output of green hydrogen is currently 0.1 million tonnes per year (International Energy Agency, 2023b).

25 Economy sizes: World Bank (2023). Germany consumes 38.1 megawatt-hours per person; France consumes 37.2 megawatt-hours per person (Energy Institute, 2024).
26 Electricity: Ember (2024). Energy: Energy Institute (2024).
27 Oil prices: Macrotrends (2024). Energy: Energy Institute (2024).
28 International Atomic Energy Agency (2024a).
29 Grubler (2010).
30 Energy supply: Energy Institute (2024). Electricity supply: Ember (2024). French reactors: International Atomic Energy Agency (2024a).
31 Ember (2024).
32 Delays: BBC News (2023d). Nuclear generation: Ember (2024).
33 Ember (2024).
34 World Nuclear News (2022a).
35 New reactors: World Nuclear News (2022a). Alliance: Euractiv (2023b). Finance Minister quote: Reuters (2023b).
36 French reactor ages: International Atomic Energy Agency (2024a). Global reactor ages: Schneider and Froggatt (2022). French reactor life extension: World Nuclear News (2022a).
37 Grubler (2010) estimated the cost of France's nuclear programme to be €230 billion (in 2008 Euros), which I inflation-adjusted to 2017 prices. For nuclear electricity increase, Ember's dataset only goes back as far as 2000, and so I multiplied France's total nuclear energy consumption in 1973 (42 terawatt-hours; Energy Institute, 2024) by France's average energy-to-electricity efficiency (38 per cent) to get 15.8 terawatt-hours of electricity; France's nuclear electricity generation in the mid-2000s was 450 terawatt-hours (Ember, 2024), which represents an increase of 434 terawatt-hours since the early 1970s. Unnerstall (2017) estimated the cost of *Energiewende*. Germany's renewable increase: Ember (2024).
38 Annual wind and solar generation increased by 2,800 terawatt-hours between 2015 and 2023 (Ember, 2024). To increase nuclear generation by that amount using Olkiluoto-3-style reactors (12 terawatt-hours each annually; see footnote 3) would require 233 of them. At $12.4 billion each (Schneider and Froggatt, 2023), it would cost $2.9 trillion in total. That's $1.2 trillion less than the world invested in renewable energy between 2015 and 2023 ($4.1 trillion; International Energy Agency, 2024).
39 This is true with and without taxes and levies (Eurostat, 2024b).
40 United Nations (2022a).
41 Lazard (2023).

42 International Renewable Energy Agency (2024).
43 Estimating the cost of electricity isn't a simple exercise. The best researchers can do is make estimates, which is why they vary from source to source. A compiling of estimates from several sources reveals the overall picture. Onshore wind and solar levelised costs: Lazard (2023), International Renewable Energy Agency (2022), and Idel (2022). Offshore wind levelised cost: Lazard (2023) and International Renewable Energy Agency (2022). Nuclear levelised cost: Lazard (2023) and Idel (2022). Onshore wind, solar, and nuclear levelised system costs: Idel (2022) and Matsuo (2022). Offshore wind levelised system cost: Matsuo (2022).
44 Reichenberg et al. (2018) estimate that the levelised system costs of wind and solar increase steadily as their share in the electricity mix increases from 20 per cent to 80 per cent; after that, their system costs increase sharply.
45 Bloomberg (2023b).
46 International Energy Agency and OECD Nuclear Energy Agency (2020).
47 Plotted using data in the previous footnote.
48 Olkiluoto-3: Schneider and Froggatt (2019). '. . . 65 to 85 per cent . . .': World Nuclear Association (2024b).
49 Reuters (2023c).
50 Fund: World Nuclear Association (2017). Surcharges: World Nuclear Association (2020c).
51 Data are for the 61 reactors connected to the grid since 2014 (International Atomic Energy Agency, 2024a).
52 Hinkley Point C: World Nuclear News (2024a). Flamanville-3: Schneider and Froggatt (2023). Analysis: Portugal-Pereira et al. (2018).
53 Hinkley Point C: World Nuclear News (2024a). Flamanville-3: American Nuclear Society (2022).
54 Flyvbjerg (2013).
55 International Atomic Energy Agency (2024a).
56 Built times: International Atomic Energy Agency (2024a). Standardised design: Grubler (2010).
57 Boiteux (2009).
58 International Atomic Energy Agency (2024a).
59 Build times: International Atomic Energy Agency (2024a). Electricity mix: Ember (2024).
60 Lovering et al. (2016).

61 Build time: International Atomic Energy Agency (2024a). Electricity mix: Emirates Nuclear Energy Corporation (2024). New power station: Reuters (2024).
62 Seagreen has a capacity of 1,075 megawatts. Assuming it has a capacity factor the same the UK offshore wind average (37.5 per cent; Department for Energy Security and Net Zero, 2024c), it has an average generation of 400 megawatts. It took 40 months to build (SSE 2023).
63 International Atomic Energy Agency (2024a).
64 British government: Department for Business, Energy, and Industrial Strategy (2023). Ten Point Plan: UK Government (2020). France: International Energy Agency (2022b). Belgium investment: World Nuclear News (2022b). Belgium 2035: World Nuclear Association (2023a).
65 Rolls-Royce SMR (nd).
66 Six football pitches: International Atomic Energy Agency (2019a). Ninety football pitches: Department of Energy & Climate Change (2013).
67 Cardboard and saltworks: The Royal Society (2020). Warming homes: Handl (1998). Chemical plant: World Nuclear News (2024b). Shut down: International Atomic Energy Agency (2024a).
68 My calculation assumes the 470 megawatt reactors will have a capacity factor of 86 per cent, our OECD electricity consumption yardstick (see page X), and a total UK electricity consumption in line with the Committee on Climate Change's 'balanced pathway' estimate (610 terawatt-hours; Committee on Climate Change, 2020).
69 Build time: International Atomic Energy Agency (2019a). Cost: World Nuclear News (2019).
70 International Atomic Energy Agency (2022a).
71 China: World Nuclear News (2022c).
72 Lawrence Livermore National Laboratory (2022a).
73 Lawrence Livermore National Laboratory (nd).
74 International Atomic Energy Agency (2022b).
75 310 watt-hours: Lawrence Livermoor National Laboratory (2022). 100,000 watt-hours: Lawrence Livermore National Laboratory (nd).
76 'Transformative moment': The Atlantic (2022). '... bountiful energy.': New York Times (2022). '... near-limitless ...': BBC News (2022c). '... precipice ...': Lawrence Livermore National Laboratory (2022b).
77 Fridge: assuming a fridge runs on 300 watts. Tesla: A Tesla Model S does 3.3 miles to the kilowatt-hour (Tesla, 2024).

78 US Atomic Energy Commission (1958).
79 US Department of Energy (2023).
80 Energy Institute (2024).
81 Kerry: The Japan Times (2023). Macron: Euractiv (2023c). '...second-biggest...': Friedlingstein *et al.* (2023). Reaffirmation: U.S. Department of Energy (2024).
82 World energy: Energy Institute (2024).
83 Microsoft: BBC News (2024a). Google: BBC News (2024b). Amazon: Financial Times (2024).
84 Hickman *et al.* (2021).
85 Guardian News (2023).

Chapter 5. The Green Sheen

1 Guardian News (2019).
2 Crew: The Guardian (2019). '... 15 days ...': BBC News (2019). Herrmann quote: The Washington Post (2019).
3 '... Herrmann flew...': France 24 (2019). '... five skippers flew...': Taz (2019). '... another skipper...': The Times (2019).
4 Bag footprint: The Danish Environmental Protection Agency (2018).
5 Guterres quote: United Nations Climate Change (2022).
6 Offshore wind data: Ritchie (2022a). All other data: United Nations Economic Commission for Europe (2022).
7 Exclusion zone areas and effect on electricity concentration: Lovering *et al.* (2022). Chernobyl wildlife: Deryabina *et al.* (2015). Fukushima wildlife: National Geographic (2021). Ecological impacts: United Nations Economic Commission for Europe (2022).
8 The six reactors together generate 34.1 terawatt-hours of electricity annually on average (International Atomic Energy Agency, 2024a) and London consumes 34.5 terawatt-hours annually (Department for Energy Security and Net Zero, 2024d). I measured the area of Gravelines Nuclear Power Station on Google Maps.
9 Assuming a capacity factor of 86 per cent, a 440 megawatt Rolls-Royce small modular reactor would generate 3,300 gigawatt-hours of electricity annually; they're designed to fit on a 40 thousand square metre site (International Atomic Energy Agency, 2019a), about the same as 6 football pitches. Leeds and Glasgow each consume 2,400 gigawatt-hours of electricity annually (Department for Energy Security and Net Zero, 2024d).

Notes 329

10 '. . . 40,000 . . .': calculated by dividing the solar farm's annual generation (Juwi, nd) by 8 megawatt-hours. '. . . 1.7 *million* OECD citizens . . .': calculated by multiplying the solar farm's area (Juwi, nd) by nuclear's electricity concentration (3,000 gigawatt-hours annually per square kilometre; United Nations Economic Commission for Europe, 2022) – which yields 13.5 million megawatt-hours – and dividing that by our 8 megawatt-hour per year OECD yardstick (see page X).
11 National Geographic (2022).
12 Reindeer: Mining Technology (2019). Ore: Reuters (2019). Lundberg: The Independent (2019).
13 International Energy Agency (2022c).
14 The World Bank (2020).
15 Wang *et al.* (2024).
16 Plotted using data from the previous footnote. 'Nuclear' on this plot is a 1.6-gigawatt European Power Reactor, the same type of nuclear reactor as Olkiluoto-3. Estimates of the mineral- and mining-intensity of different power sources vary, but the data give the same overall picture: nuclear is the least mining-intensive source of power, and fossil fuels are by far the most intensive.
17 Cobalt supply: International Energy Agency (2022c). List: US Deptartment of Labor (2022). Demand increase: The World Bank (2020).
18 Mineral extraction and processing: International Energy Agency (2022c).
19 Stop Sizewell C (2021).
20 CO_2: The Planning Inspectorate (2021). My calculation assumes a capacity factor of 86 per cent and a lifetime of 60 years.
21 Schlömer *et al.* (2014).
22 Plotted using data from Schlömer *et al.* (2014).
23 Law: European Parliament (2023). Exemptions and subsidies: Harrison (2021). Declaration: Reuters (2017).
24 Carbon dioxide: Sterman *et al.* (2018). Seven million tonnes: Schlesinger (2018).
25 Schlesinger (2018).
26 Sterman *et al.* (2018).
27 Schlömer *et al.* (2014).
28 Sterman *et al.* (2018).
29 UK's most intense: Mayo (2024). European ranking: Harrison and Fox (2023). Carbon budget: BBC News (2024c).
30 BBC News (2024c).
31 MacDonald (2022).

Chapter 6. Nuclear Waste

1. Savanta and Radiant Energy Group (2023).
2. Rutherford (1900).
3. Rutherford (1900).
4. Half-lives – here and henceforth – from Kondev et al. (2021).
5. Radioactivity: World Nuclear Association (2022c). Volume: International Atomic Energy Agency (2022c).
6. Choppin et al. (2013a).
7. Choppin et al. (2013a).
8. McMillan (1939).
9. McMillan (1951).
10. Meitner et al. (1937).
11. McMillan and Ableson (1940).
12. Ojovan et al. (2019).
13. In my calculation, I assume a 1,000-megawatt reactor with an 86 per cent capacity factor (see footnote 3, Chapter 4) goes through 30 tonnes of fuel per year (World Nuclear Association, nd), and that spent fuel has a density the same as uranium dioxide (11 grams per cubic centimetre). That yields 230 cubic centimetres of spent fuel over an average OECD 80-year lifetime, which is smaller than the glasses of wine they serve in my local.
14. Global spent fuel mass: International Atomic Energy Agency (2020b). I assumed it has a density the same as uranium dioxide (11 grams per cubic centimetre).
15. Assuming the ~ 400 reactors in the global fleet each generate 30 tonnes of spent fuel per year.
16. BBC Future (2023).
17. Trauth et al. (1993).
18. Total electricity: International Atomic Energy Agency (2024a). Electricity mix: Ember (2024). High-level waste: International Atomic Energy Agency (2023c). Nuclear Energy Act: Finlex (2020). Approval: Posiva (2023).
19. In 2023, Finland consumed 79.8 terawatt-hours of electricity in 2023 (Ember, 2024) and 24.7 terawatt-hours came from Olkiluoto (International Atomic Energy Agency, 2024a).
20. Age: Geological Survey of Finland (2023). Geology: Aaltonen et al. (2016).
21. Posiva (nd).
22. Radioactivity and walls: Posiva (nd). Encapsulation: Posiva Oy (2023).

23 Bundling: Posiva Oy (2023). Cylinders: Posiva (nd). Copper: Salonen *et al.* (2021).
24 Posiva Oy (2023).
25 Posiva (nd).
26 El-Showk (2022).
27 Posiva (nd).
28 Full, clay, and demolition: Posiva (nd). '. . . 6,000 tonnes . . .': BBC Focus (2023). '. . . 12,000 tonnes . . .': World Nuclear Association (2024c).
29 El-Showk (2022).
30 Radiation and Nuclear Safety Authority (2015).
31 Onkalo cost: World Nuclear Association (2024c). Globally: World Nuclear Association (2024d).
32 Finnish Energy (2022).
33 Sweden: American Nuclear Society (2024a). France: Lehtonen (2023). Other nations: World Nuclear Association (2024d).

Chapter 7. Nuclear for the Third Millennium

1 DelCul and Spencer (2020).
2 Spent fuel: International Atomic Energy Agency (2019b). Recycling: International Atomic Energy Agency (nd,b).
3 Glass: Orano (nd,a).
4 Amount and energy: Orano (2022). I used the energy conversion factors from International Energy Agency (2023d).
5 Orano (2022).
6 International Atomic Energy Agency (2007).
7 Fan *et al.* (2010).
8 Calculated by dividing the mass of France's recycled uranium stockpile (34,000 tonnes; Orano, 2022) by a heavy-water reactor's recycled uranium consumption rate (50 tonnes per year; International Atomic Energy Agency, 2007).
9 Ronen (2006).
10 Half-lives – here and henceforth – from Kondev *et al.* (2021).
11 Mass: International Atomic Energy Agency (2023d). Largest: Bodel *et al.* (2023). Timescale: Ojovan *et al.* (2019). Volume calculation is based on a plutonium dioxide density of 11.5 tonnes per cubic metre.
12 Hyatt (2020) estimates the UK's plutonium could power a pair of 1.6 gigawatt Hinkley Point C-type reactor for 90 years. My calculation assumes

a capacity factor of 86 per cent (see footnote 3, Chapter 4) and an average OECD electricity consumption (see page X).

13 Department for Energy Security and Net Zero (2024e).
14 Britain: Hyatt (2017). France: Grenèche (2002).
15 Carbajo (2001).
16 Orano (nd,b).
17 Orano (nd,c).
18 30 per cent: OECD/NEA (2021).
19 France generated 15,700 terawatt-hours of nuclear electricity between 1965 and 2023; the UK generated 3,500 terawatt-hours (Energy Institute, 2024). As of 2023, France's stockpile is 106 tonnes (International Atomic Energy Agency, 2023e); the UK's is 141 tonnes (International Atomic Energy Agency, 2023d).
20 International Atomic Energy Agency (2012a).
21 Odds of splitting taken from the JEFF-3.3 nuclear data library (Plompen et al. 2020). Neutron numbers: Duderstadt and Hamilton (1976b).
22 Total fast reactors and accumulated runtime: World Nuclear Association (2021a). Commercial operation: International Atomic Energy Agency (2024a).
23 International Atomic Energy Agency (2012a).
24 OECD/NEA (2021).
25 Duderstadt and Hamilton (1976b).
26 US Atomic Energy Commission (1944).
27 My calculation uses the heat values from World Nuclear Association (2020a), a thermal energy-to-electrical energy efficiency of 33 per cent (World Nuclear Association, 2020b), and a typical OECD annual electricity consumption (see page X).
28 Number of breeders: Bodel et al. (2023). Depleted uranium: Nuclear Decommissioning Authority (2023).
29 Fermi: Mitchell III and Turner (1971). Timescale: International Atomic Energy Agency (2012a). First: Cochran et al. (2010).
30 Fuel: Cochran et al. (2010). Design: Kittel et al. (1957).
31 Idaho National Engineering Laboratory (1979).
32 Kittel et al. (1957).
33 International Atomic Energy Agency (2012a).
34 Accident: McLaughlin et al. (2000a). Switched off: Cochran et al. (2010).
35 Thorium-232 abundance: Meija et al. (2016).
36 Nath (2022).

37 Mahanti (2007).
38 Nath (2022).
39 Nath (2022).
40 Wadia (2015).
41 Bhabha (1948).
42 Woddi *et al.* (2009).
43 Nath (2022).
44 International Atomic Energy Agency (2024a).
45 Nath (2022).
46 Nath (2022).
47 Woddi *et al.* (2009).
48 L'Annunziata (2016).
49 Cochran *et al.* (2010).
50 Seaborg: Mitchell III and Turner (1971). Projection: Skjoeldebrand (1973).
51 There were 350 gigawatts of nuclear capacity in the year 2000 (International Atomic Energy Agency, 2001), of which breeder reactors contributed 0.8 gigawatts (International Atomic Energy Agency, 2024a).
52 Cochran *et al.* (2010).
53 International Atomic Energy Agency (2024a).
54 Mid-1980s: Woddi *et al.* (2009). Electricity mix: Ember (2024).
55 Agency: International Atomic Energy Agency (2022d). India: World Nuclear News (2024c). TerraPower: World Nuclear News (2024d).
56 UN. Secretary-General (1987).
57 I calculated these back-of-the-envelope timescales based on the following numbers: current world energy consumption from fossil fuels (140,000 terawatt-hours per year; Energy Institute, 2024); known uranium reserves (7.9 million tonnes; Nuclear Energy Agency and the International Atomic Energy Agency, 2023); world stockpile of depleted uranium (1.2 million tonnes; OECD Nuclear Energy Agency and International Atomic Energy Agency, 2001; note that this estimate dates from 1999, and is therefore conservative); spent fuel that hasn't been chemically separated, which I assume is 96 per cent uranium and plutonium (253,000 tonnes; International Atomic Energy Agency, 2022c); known thorium reserves (6.2 million tonnes; International Atomic Energy Agency, 2019c); energy-densities for natural uranium in a light-water reactor (139 megawatt-hours per kilogram), natural uranium in a light-water reactor with MOX recycling (181 megawatt-hours per kilogram),

and natural uranium in a breeder reactor (7,780 megawatt-hours per kilogram), from World Nuclear Association (2020a); and I assumed thorium, depleted uranium, and spent fuel in a breeder reactor has the same energy-density as natural uranium.

58 Oceanic uranium: Liu *et al.* (2017). Concentration: Tsouris *et al.* (2017).

Chapter 8. Radiophobia

1 United Nations Scientific Committee on the Effects of Atomic Radiation (2008a).
2 United States Nuclear Regulatory Commission (2021).
3 United Nations Scientific Committee on the Effects of Atomic Radiation (2008a).
4 United Nations Scientific Committee on the Effects of Atomic Radiation (2008a).
5 Ministry of Defence (2020).
6 United Nations Scientific Committee on the Effects of Atomic Radiation (2008a).
7 Tap water and ingestion dose: United Nations Scientific Committee on the Effects of Atomic Radiation (2008a). Brazil nuts: Martins *et al.* (2012).
8 Half-lives: Kondev *et al.* (2021). Dose: United Nations Scientific Committee on the Effects of Atomic Radiation (2008a).
9 World data: United Nations Scientific Committee on the Effects of Atomic Radiation (2008a). Europe: European Commission Joint Research Centre (2019).
10 UK Health Security Agency (nd).
11 State doses: Mauro *et al.* (2005). State cancers: National Cancer Institute (2020).
12 Nair *et al.* (2009).
13 Mortazavi *et al.* 2019.
14 Fallout dose: United Nations Scientific Committee on the Effects of Atomic Radiation (2008a).
15 Fallout extent: International Atomic Energy Agency (nd,c). Chernobyl dose: United Nations Scientific Committee on the Effects of Atomic Radiation (2008a).
16 United Nations Scientific Committee on the Effects of Atomic Radiation (2008a).
17 United Nations Economic Commission for Europe (2022).
18 Eve (1939).

19 Curie (1905).
20 Death: Atomic Heritage Foundation (nd). Life expectancy: Institut National d'Etudes Démographiques (2020).
21 Muller (1946).
22 Boice (2015).
23 Centres for Disease Control and Prevention (2019).
24 McLaughlin et al. (2000b), Hempelmann et al. (1979), and Hempelmann et al. (1952).
25 Centres for Disease Control and Prevention (2019).
26 Radiation Effects Research Foundation (2016).
27 Radiation Effects Research Foundation (2016).
28 Dose reconstruction: Radiation Effects Research Foundation (2016). Dose range: Ozasa et al. (2012).
29 Radiation Effects Research Foundation (2016).
30 Radiation Effects Research Foundation (2015).
31 Data from Ozasa et al. (2012).
32 Rule of thumb: International Commission on Radiological Protection (2007). EU: Eurostat (2023).
33 McLean et al. (2017).
34 US Department of Energy (2010).
35 Number: Mahesh et al. (2023). Dose: United Nations Scientific Committee on the Effects of Atomic Radiation (2008a).
36 Mettler et al. (2008).
37 Mettler et al. (2020).
38 Ministry of Defence (2020).
39 Cucinotta et al. (2008).
40 Reynolds et al. (2019).

Chapter 9. We Need To Talk About Chernobyl

1 Sveriges Radio (2019).
2 Reactors: International Atomic Energy Agency (2024a). Poll: Radiant Energy Group (2023).
3 Joyce (2018).
4 World Nuclear Association (2022d).
5 Chernobyl: World Nuclear Association (2022e). Northern hemisphere: United Nations Scientific Committee on the Effects of Atomic Radiation (2008a).

6 United Nations Scientific Committee on the Effects of Atomic Radiation (2008a).
7 Mettler *et al.* (2007).
8 United Nations Scientific Committee on the Effects of Atomic Radiation (2008b).
9 Workplace deaths: Health and Safety Executive (nd).
10 Timescale: United Nations Scientific Committee on the Effects of Atomic Radiation (2008b). Health outcomes: Rahu *et al.* (2023).
11 '... comparable to ...': United Nations Scientific Committee on the Effects of Atomic Radiation (2008b). Average dose: United Nations Scientific Committee on the Effects of Atomic Radiation (2008a). 2019 study: Leung *et al.* (2019).
12 United Nations Scientific Committee on the Effects of Atomic Radiation (2008b).
13 Half-lives – here and henceforth – from Kondev *et al.* (2021). Response: OECD and Nuclear Energy Agency (2002). Milk: United Nations Scientific Committee on the Effects of Atomic Radiation (2008b).
14 Half-lives – here and henceforth – from Kondev *et al.* (2021). Response: OECD and Nuclear Energy Agency (2002). Milk: United Nations Scientific Committee on the Effects of Atomic Radiation (2008b).
15 Cases and estimate: United Nations Scientific Committee on the Effects of Atomic Radiation (2018). Survival rate: Zhang *et al.* (2022).
16 Union of Concerned Scientists: Grolund (2011). Series: HBO (2019).
17 Veritasium (2016).
18 Stigma and mental health: Bennett *et al.* (2006). Quote: World Health Organization (2020). Mothers: Bromet *et al.* (2011).
19 Abortions: Ketchum (1987). Denmark: Knudsen (1991). Italy: Spinelli and Osborn (1991). Greece: Trichopoulos et al. (1987). Effects: Bennett *et al.* (2006). Quote: New York Times (1987).
20 Liquidator mental health: Laidra *et al.* (2017). Estonian liquidators: Rahu *et al.* (2023).
21 Turmoil: Bromet *et al.* (2011). Exacerbation: Bennett *et al.* (2006).
22 Honshū: NASA Earth Observatory (2021). Most powerful: U.S. Geological Survey (2024).
23 National Research Council U.S. (2014).
24 Melted: World Nuclear Association (2024e). Chemical reaction: National Research Council U.S. (2014).

25 Units 1 and 3: National Research Council U.S. (2014). Unit 4: World Nuclear Association (2024e).
26 World Nuclear Association (2024e).
27 United Nations Scientific Committee on the Effects of Atomic Radiation (2013).
28 Death: BBC News (2018). Doses: United Nations Scientific Committee on the Effects of Atomic Radiation (2022a).
29 United Nations Scientific Committee on the Effects of Atomic Radiation (2022a).
30 Fukushima: United Nations Scientific Committee on the Effects of Atomic Radiation (2022a). Colorado: Mauro *et al.* 2005.
31 Guidelines: International Commission on Radiological Protection (2007). Admission: International Atomic Energy Agency (nd,d). Regarding evacuation, I'm referring to US states (Mauro *et al.* 2005) and European nations (European Commission, Joint Research Centre, 2019) with annual background radiation doses 1 millisievert over the global average of 2.4 millisieverts (United Nations Scientific Committee on the Effects of Atomic Radiation, 2008a).
32 Evacuation numbers: Fukushima Prefecture and International Atomic Energy Agency (2021). Evacuation deaths: Reconstruction Agency (2020). Elderly: World Nuclear Association (2024e).
33 Radiation levels: World Nuclear Association (2024e). Thomas quote: Financial Times (2018).
34 Contaminated earth: Normile (2016). Colorado and North Dakota: Mauro *et al.* (2005). Cornwall: UK Health Security Agency (nd). Exclusion zone: Fukushima Prefecture (2024). Evacuees: Fukushima Prefecture and International Atomic Energy Agency (2021).
35 Sutou: Sutou (2016).
36 Contaminants: International Atomic Energy Agency (nd,e). Volume: International Atomic Energy Agency (2023f). I assume an Olympic-sized swimming pool holds 2.5 million litres of water.
37 Tritium: International Atomic Energy Agency (2023f).
38 2070s: BBC News (2023e). Chinese government: Embassy of the People's Republic of China in the Kingdom of Thailand (2023). South Korea and Hong Kong: Reuters (2023d). Greenpeace: Greenpeace (2023).
39 Tritium amount and dose: International Atomic Energy Agency (2023). Drinking water: World Health Organization (2022).

40 Energy: United States Nuclear Regulatory Commission (2019). Residence time: United Nations Scientific Committee on the Effects of Atomic Radiation (2017).
41 Cosmogenic tritium: Choppin *et al.* (2013b). About 280 grams of tritium are produced in the sky every year (International Atomic Energy Agency, 2023).
42 'Disaster' quote: Embassy of the People's Republic of China in the Kingdom of Thailand (2023). 'Selfish' quote: BBC News (2023f). Japan tritium: International Atomic Energy Agency (2023f). China tritium: China Nuclear Energy Association (2021).
43 France tritium 2017–2019: Orano (2020). France tritium 2020–2022: Orano (2023) France tritium 2023: Orano (2024). Ecosystem and dose: Smith *et al.* (2023).
44 Sellafield Ltd. (2023).
45 Plotted using data from Ritchie (2022b).
46 USA reactors: International Atomic Energy Agency (2024a). Deaths (or lack thereof): United Nations Scientific Committee on the Effects of Atomic Radiation (2008b).
47 Rain and deaths: Sovacool *et al.* (2016). Wall of water: Yang *et al.* (2017).
48 Sovacool *et al.* (2016).
49 Air pollution: Lelieveld *et al.* (2019). Natural disasters: Ritchie and Rosado (2022). HIV/AIDS: World Health Organization (2023). COVID-19: World Health Organization (2024). Ranking: Institute for Health Metrics and Evaluation (2019).
50 Lelieveld *et al.* (2019).
51 Kharecha and Hansen (2013).
52 RBMK statistics: International Atomic Energy Agency (2024a).
53 Stoner *et al.* (2021).
54 Ember (2024).
55 Japan deaths: Kharecha and Sato (2019). Germany deaths: Jarvis *et al.* (2022).
56 Neidell *et al.* (2021).
57 Reactors: International Atomic Energy Agency (2024a). Electricity mix: Ember (2024).
58 Kajiyama: Financial Times (2021). Aim: Agency for Natural Resources and Energy (2021). Declaration: US Department of Energy (2023).

Chapter 10. Golden Geese

1 International Atomic Energy Agency (nd,f).
2 U.S. Food & Drug Administration (2016).
3 Recycling: Organisation for Economic Co-operation and Development (2022). Plastics and radiation: Ponomarev *et al.* 2022.
4 Schneider *et al.* (2001)
5 Radiation exposure: O'Donnell *et al.* (1981). Woman who swallowed americium: Rundo *et al.* (1977).
6 Kondev *et al.* (2021).
7 International Atomic Energy Agency (nd,g).
8 Chinese rice: International Atomic Energy Agency (2022e). Japanese tomatoes: International Atomic Energy Agency (2022f). Mauritian oyster mushrooms: International Atomic Energy Agency (2022g).
9 American roses: International Atomic Energy Agency (2022h). Dutch tulips: International Atomic Energy Agency (2022i). Indian gladioli: International Atomic Energy Agency (2022j).
10 Curry (2016).
11 Soddy (1920). Ironically, *not* thawing the frozen poles is one of the appeals of nuclear technology.
12 Johnson (2012).
13 Atomic Energy Association of Great Britain and Ladies' Atomic Energy Club: Johnson (2012). Muriel quotes: Time (1950).
14 Time (1950).
15 Johnson (2012).
16 *UNA-La Molina 95*: International Atomic Energy Agency (2022k). *Centenario II* the 2006 Peruvian Prize of Good Governmental Practices: International Atomic Energy Agency (2012b).
17 *Binadhan-7*: International Atomic Energy Agency (2022l). Suruj Ali: International Atomic Energy Agency (2017).
18 Tran *et al.* (2006).
19 United States Department of Agriculture (2018).
20 Reichard (nd).
21 Klassen *et al.* (2021).
22 The Golden Goose Award (2016).
23 Diets: Bushland and Hopkins (1953). Experiment: see Klassen *et al.* (2021).
24 Baumhover (2002).

25 Goats: World Food Prize Foundation (nd). Curaçao: Baumhover (2002).
26 Florida: Klassen *et al.* (2021). Production rate: Baumhover (2002).
27 Klassen *et al.* (2021).
28 Diet: National Geographic (2019). Drop rate: Centro de Dispersión de Moscas Estériles (nd).
29 Lindquist *et al.* (1992).
30 Flies: International Atomic Energy Agency (nd,h). Moths: International Atomic Energy Agency (nd,i).
31 Dengue: World Health Organization (2019). Malaria: International Atomic Energy Agency (nd,j). Death toll: see Rosser (2022).
32 Fauna & Flora International (nd,a).
33 White rhinos: Ritchie (2022c). Vince: BBC News (2017). Black rhinos WWF (nd,a). Javan rhinos: WWF (nd,b).
34 Bogus horn properties: WWF (nd,c). Monetary value: BBC News (2015a).
35 '... armed guards ...': Forbes (2024). Poaching tactics: WWF (nd,d).
36 I interviewed Larkin over FaceTime in February 2024, and he told me the backstory of *Rhisotopes*.
37 It's fun to calculate the radioactivity of a banana. Worked example: Ball *et al.* (2004).
38 Fauna & Flora International (nd,b).
39 World Food Prize: World Food Prize Foundation (nd). 'Their own words': @GoldGooseAward (nd). 2016 Golden Goose Award: The Golden Goose Award (2016).

Chapter 11. Radioactive Remedies

1 Riesz (1995).
2 Röntgen (1896).
3 Half-lives – here and henceforth – from Kondev *et al.* (2021).
4 Dose: International Atomic Energy Agency (2024b). Cornwall: UK Health Security Agency (2023).
5 Deaths: World Health Organization (2022). 'On the rise': United Nations (2024a).
6 United Nations Scientific Committee on the Effects of Atomic Radiation (2022b).
7 United Nations Scientific Committee on the Effects of Atomic Radiation (2018).

8 Surgery: Mayo Clinic (2024a).
9 Therapy: Mayo Clinic (2024b). Dose: Bennett *et al.* (2016).
10 Airports: Gangopadhyay *et al.* (2006).
11 Grigsby *et al.* (2000).
12 Mayo Clinic (2024b).
13 Strontium-89: Guerra Liberal *et al.* (2016).
14 '... most prevalent ...': Wang *et al.* (2022). '... second only ...': Luo *et al.* (2022). '... long-term ...': Rebello *et al.* (2021). '... resistant to treatment ...': Ruiz De Porras *et* al. (2021).
15 Benešová *et al.* (2015).
16 Fallah *et al.* (2023).
17 USA: U. S. Food and Drug Administration (2022). EU: European Medicines Agency (2022). Quote: Czernin and Calais (2021).
18 Collateral: National Cancer Institute (2021). Nausea and fatigue: Fallah *et al.* 2023. New drugs: Jalilian and Albon. 2023.
19 Terbium-161: Al-Ibraheem and Scott (2023). Rhenium-188: Tietze *et al.* (2023). Scandium-47: Siwowska *et al.* 2019. Holmium-166: Klaassen *et al.* 2019. Yttrium-90: Mayo Clinic Health System (2023).
20 Distance: Pallares and Abergel (2022). Alpha and beta particles vary in their energies (depending on which isotope they originated from), but a good rule of thumb is that the former carries about five million electron-volts whilst the latter carries about half a million electron-volts.
21 Approval: Kluetz *et al.* (2014). Radium-223: Sartor *et al.* (2014).
22 Pallares *et al.* (2022).
23 Sathekge *et al.* (2024).
24 Kratochwil *et al.* (2016).
25 Leukemia: U.S. National Library of Medicine (2023a). Colon: U.S. National Library of Medicine (2024a). Eye: Eye: U.S. National Library of Medicine (2024b). White blood cells: U.S. National Library of Medicine (2024c).
26 Thyroid: U.S. National Library of Medicine (2024d). Lung: U.S. National Library of Medicine (2024e). Neuroendocrine: U.S. National Library of Medicine (2023b). Leukaemia: Rosenblat *et al.* (2010). Lymphoma: Lindén *et al.* (2021). Ovarian: U.S. National Library of Medicine (2023c). Infectious diseases: Helal and Dadachova (2018).
27 Joliot and Curie (1934).
28 Joloit (1935).
29 Rutherford (1927).
30 Lawrence quote: U.S. Department of Energy (2013).

31 Prototype: Lawrence (1951). 'By 1939 . . .': Yarris (2021). '. . . 16 *million* . . .': Siegbahn (1939).
32 '. . . 60-inch cyclotron . . .': Lawrence (1951). '. . . Seaborg and his colleagues . . .': Nobel Lectures (1964). Iodine-131: Livingood and Seaborg (1938). Seaborg's mother: Seaborg (2019).
33 Peeva (2021).
34 Cole et al. (2014).
35 Astatine-211: Albertsson et al. (2023). Actinium-225: Apostolidis et al. (2005).
36 World Nuclear Association (2021b).
37 Vogel et al. (2021).
38 Ferris et al. (2021).
39 Three dozen: Jalilian and Albon (2023). Scandium-47: Siwowska et al. (2019). Yttrium-90: Advancing Nuclear Medicine (nd). Terbium-161: Müller et al. (2019). Holmium-166: Klaassen et al. (2019). Rhenium-188: Kleynhans et al. (2023). Uranium: World Nuclear Association (2021b).
40 Alvarez (2013).
41 Mature stockpile: Shea (2018).
42 Shea (2018).
43 Yttrium-90, holmium-166, and rhenium-188: Le et al. (2014). Lead-212: World Nuclear News (2022d). Bismuth-213: Ahenkorah et al. (2021). Radium-223 and thorium-227: U.S. Department of Energy (2017).
44 2010 crisis: Vogel et al. (2021). High Flux Reactor: European Commission (2020).
45 International Atomic Energy Agency (2024c).
46 Vogel et al. (2021).
47 World Nuclear News (2022e).
48 Robertson et al. (2018).
49 Hey and Walters (1997).
50 Joloit (1935).

Chapter 12. Stockpiles and Sleuths

1 Webb (1949).
2 Giovannitti and Freed (1965).
3 Frisch (1974).
4 Estimate: Fermi (1945). Gadget's yield is surprisingly uncertain; estimates range from 18.6 to 24.8 kilotons (Selby et al. 2021).

5 Gadget's 6-kilogram plutonium core (United States Department of Energy 2002) had a density of about 15.9 grams per cubic centimetre (Hecker 2000).
6 Design: U.S. Department of Energy (nd). Fat Man: Malik (1985).
7 Design: U.S. Department of Energy (nd). Little Boy: Malik (1985).
8 Soviet Union: Bleek (2017).
9 Ivy Mike power: Miller (2022). Ivy Mike mushroom: Department of the Air Force (1985). Tsar Bomba: Atomic Heritage Foundation (2014).
10 Toon et al. (2019) modelled what would happen if India and Pakistan went to war and detonated 250 atom bombs.
11 Sagan (1983).
12 Tests: Arms Control Association (2023).
13 Dose: United Nations Scientific Committee on the Effects of Atomic Radiation (2008a). Strontium-90: Stamoulis et al. (1999). Tests: Arms Control Association (2023).
14 Soil: Meusburger et al. 2020. Carbon-14: United Nations Scientific Committee on the Effects of Atomic Radiation (2008a).
15 Hiatus: National Archives and Records Administration (nd). Tests: Arms Control Association (2023).
16 Number of committees: United Nations Office for Disarmament Affairs (2024a), with major processing from Herre et al. (2024a). Committees: United Nations Treaty Collection (nd).
17 Treaty: UN Treaty Collection (2024). Annex II States: U.S. Department of State (nd). Russia: Reuters (2023e).
18 Reactors: International Atomic Energy Agency (2024a). Stockpile: Kristensen et al. (2024). The exact sizes of atomic bomb stockpiles are state secrets; data in Kristensen et al. (2024) are estimates.
19 Reactors: International Atomic Energy Agency (2024a). Stockpile: Kristensen et al. (2024).
20 Kristensen et al. (2024).
21 Arms control: Arms Control Association (nd).
22 Suh (2023), with major processing from Herre et al. (2024a).
23 Military aircraft: Kratz (2021). Norwegian rocket: Atomic Heritage Foundation (2018).
24 Bleek (2017).
25 Bomb acquisition: Bleek (2017). Non-Proliferation Treaty: United Nations Office for Disarmament Affairs (2024b). World Health Organization: United Nations (2024b). Charter: United Nations (2024c).

26 Treaty: United Nations (1968). South Africa: von Baeckmann et al. (1995).
27 Foreign, Commonwealth & Development Office (2023).
28 Achilles' heel: ElBaradei (2009).
29 International Atomic Energy Agency (2023g).
30 Bleek (2017).
31 Pavlov and Rybachenkov (2013).
32 Neff (1991).
33 Deal: US Government (1993).
34 Calculated by multiplying the USA's total electricity consumption between 1993 and 2013 (85 million gigawatt-hours; Energy Institute, 2024) by the percentage it derived from Megatons to Megawatts (10 per cent; American Institute of Physics 2020). I assumed a CO_2 intensity of 490 grams per kilowatt-hour for gas and 820 grams per kilowatt-hour for coal, which are the global medians (Schlömer et al. 2014). Emissions from world regions from Friedlingstein et al. (2023b).
35 American Academy of Achievement (2016).
36 American Academy of Achievement (2016).
37 Franck (1945).
38 American Academy of Achievement (2016).
39 Reagan and Gorbachev: National Archives (1985). '. . . in 2022 . . .': The White House (2022).
40 War occurrence: Lyall (2020). Relationships: Diehl et al. (2023). Armed conflict deaths: Uppsala Conflict Data Program (2023) and Peace Research Institute Oslo (2017). Democracy: Coppedge et al. (2024). All these datasets are compiled, processed, and presented by Herre et al. (2024b).
41 United Nations (2022b).
42 Nathwani et al. (2016).
43 Gamma-rays: BBC News (2015b). Dose: Owen (2016).
44 Lethal dose: Harrison et al. (2007).
45 er KBG: BBC News (2016). Bar: Owen (2016).
46 Owen (2016).
47 Lead: Black (2012). Failed attempt: Owen (2016).
48 Mayak: Dombey (2007).
49 British: Owen (2016). European: European Court of Human Rights (2021). Denial: BBC News (2016).
50 International Atomic Energy Agency (2024d).
51 STO (nd). Half-lives – here and henceforth – from Kondev et al. (2021).

52 BBC News (2023g).
53 Sickness: BBC News (2023g). Timeline: BBC News (2023h).
54 Find: ANSTO (nd). Marks: BBC News (2023h).
55 Shape of the bomb pulse from Hua *et al.* (2022).
56 Fournier and Ross (2013).
57 Spalding *et al.* (2005).
58 Tsunami: Hopkin (2005). Italian corpse: Calcagnile *et al.* (2013).
59 Grimm (2008).
60 Origin: Wallenius *et al.* (2020). Remit: Nuclear Forensics International Technical Working Group (2024). Numbers: U.S. Department of State (2022).

Chapter 13. The Final Frontier

1 NASA (nd).
2 Jupiter: Uri (2019). Saturn: Uri (2020).
3 Plutonium-238 has a heat output of 0.56 watts per gram (Dustin and Borrelli, 2021) and plutonium dioxide has a density of 11.5 grams per cubic centimetre. I assumed a tealight has a heat output of 30 watts.
4 Lee and Bairstow (2015).
5 Battery: Lee and Bairstow (2015). Half-life: Kondev *et al.* (2021).
6 Lindblom (2018).
7 Forecast: Hess *et al.* (1976).
8 World Nuclear Association (2023b).
9 Nuclear Decommissioning Authority (2008).
10 I assumed that the UK's 141-tonne stockpile (International Atomic Energy Agency, 2022m) is somewhere between 5 and 15 per cent plutonium-241 and is 30 years old on average.
11 Purification: Brown *et al.* (2018). Rover: Gibney (2024).
12 Temperature: Stopar (nd). Cabeus: Luchsinger *et all* (2021). Water: Gasparini and Wasser (nd).
13 Seaborg (1968).
14 SNAP-10A: Johnson *et al.* (1967). Orbit: Voss (1984).
15 International Atomic Energy Agency (2005).
16 Energy For Space: US Department of Energy (2021). Russia and China: American Nuclear Society (2024b). Rolls-Royce: UK Space Agency (2024).
17 Dose estimate: Zeitlin *et al.* (2013). Trip time: NASA (2021).

18 Saturn V fuel: NASA (1968). For my calculation, I assumed chemical fuel has an energy-density of 50 million joules per kilogram and 90 per cent enriched uranium has an energy-density of 74 trillion joules per kilogram.
19 Journey time: Emrich (2022).
20 Hula (2015).
21 Hall (2023).
22 Jet Propulsion Laboratory (nd,a).
23 Future: Jet Propulsion Laboratory (nd,b). Far future: Jet Propulsion Laboratory (nd,a).

Epilogue: Appealing to Our Better Nature

1 Roser (2023b).

Index

16 pp To Come

Dr Tim Gregory is a nuclear chemist at the United Kingdom National Nuclear Laboratory at Sellafield, public speaker, broadcaster, and author of *Meteorite: How Stones from Outer Space Made our World*. He has a PhD from the University of Bristol and lives in the North of England. His website is HYPERLINK "http://www.tim-gregory.co.uk"www.tim-gregory.co.uk.